Urban Health and Wellbeing

Systems Approaches

Series editor

Yong-Guan Zhu, Xiamen, China

The Urban Health and Wellbeing: Systems Approaches series is based on a 10-year global interdisciplinary research program developed by International Council for Science (ICSU), and sponsored by the InterAcademy Partnership (IAP) and the United Nations University (UNU). It addresses up-to-date urban health issues from around the world and provides an appealing integrated urban development approach from a systems perspective. This series aims to propose a new conceptual framework for considering the multi-factorial and cross sectorial nature of both determinants and drivers of health and wellbeing in urban populations and takes a systems approach for improving the understanding of the interconnected nature of health in cities. The systems approach includes an engagement with urban communities in the process of creating and transferring knowledge. Further, it aims at generating knowledge and providing the evidence that is relevant to people and policy-makers for improving integrated decision making and governance for the health and wellbeing of urban dwellers. The methods applied, come from various epistemological domains in order to improve understanding of how the composition and functioning of urban environments impacts physical, mental and social heath and how inequalities can be reduced to improve the overall quality of urban life.

The systems approach is applied to science and society and defined by a deep investigation into disciplinary knowledge domains relevant for urban health and wellbeing, as well as an inter- and transdisciplinary dialogue and shared understanding of the issues between scientific communities, policy makers and societal stakeholders more broadly. It involves one or more of the following elements: (1) the development of new conceptual models that incorporate dynamic relations among variables which define urban health and wellbeing; (2) the use of systems tools, stimulation models and collaborative modelling methods; (3) the integration of various sources and types of data including spatial, visual, quantitative and qualitative data.

Like the first book, the coming books will all address the topic of urban health and wellbeing, specifically by taking a systems approach. The topics range across all urban sectors and can, for example, cover the following areas:

(1) transportation, urban planning and housing, urban water, energy and food, communication, resources and energy, urban food systems, public service provision, etc.
(2) the related health disorders in physical, social and mental health
(3) the methods and models used and the type of science applied to understand the complexity of urban health and wellbeing.

More information about this series at http://www.springer.com/series/15601

Franz W. Gatzweiler · Yong-Guan Zhu
Anna V. Diez Roux · Anthony Capon
Christel Donnelly · Gérard Salem
Hany M. Ayad · Ilene Speizer
Indira Nath · Jo I. Boufford
Keisuke Hanaki · Luuk C. Rietveld
Pierre Ritchie · Saroj Jayasinghe
Susan Parnell · Yi Zhang

Advancing Health and Wellbeing in the Changing Urban Environment

Implementing a Systems Approach

ZHEJIANG UNIVERSITY PRESS
浙江大学出版社

Springer

Franz W. Gatzweiler
Institute of Urban Environment
Chinese Academy of Sciences
Xiamen, Fujian
China

Yong-Guan Zhu
Institute of Urban Environment
Chinese Academy of Sciences
Xiamen, Fujian
China

Anna V. Diez Roux
School of Public Health
Drexel University
Philadelphia, PA
USA

Anthony Capon
International Institute for Global Health
United Nations University
Kuala Lumpur
Malaysia

Christel Donnelly
Department of Infectious Disease Epidemiology
Imperial College London
London
UK

Gérard Salem
Université Paris-Ouest
Nanterre
France

Hany M. Ayad
Faculty of Engineering
Alexandria University
Alexandria
Egypt

Ilene Speizer
Department of Maternal and Child Health
Carolina Population Center
Chapel Hill, NC
USA

Indira Nath
National Institute of Pathology
New Delhi
India

Jo I. Boufford
New York Academy of Medicine
New York, NY
USA

Keisuke Hanaki
Department of Urban Engineering
The University of Tokyo
Tokoyo
Japan

Luuk C. Rietveld
Department of Water Management
Delft University of Technology
Delft
The Netherlands

Pierre Ritchie
Faculty of Social Sciences
University of Ottawa
Ottawa
Canada

Saroj Jayasinghe
Faculty of Medicine
University of Colombo
Colombo
Sri Lanka

Susan Parnell
Department of Environmental
and Geographical Sciences
University of Cape Town
Cape Town, Western Cape
South Africa

Yi Zhang
Institute of Urban Environment
Chinese Academy of Sciences
Xiamen
China

ISSN 2510-3490 ISSN 2510-3504 (electronic)
Urban Health and Wellbeing
ISBN 978-981-10-3363-6 ISBN 978-981-10-3364-3 (eBook)
DOI 10.1007/978-981-10-3364-3

Jointly published with Zhejiang University Press, Hangzhou, China

Library of Congress Control Number: 2017932008

Printed on acid-free paper

This Springer imprint is published by Springer Nature
The registered company is Springer Nature Singapore Pte Ltd.
The registered company address is: 152 Beach Road, #21-01/04 Gateway East, Singapore 189721, Singapore

URBAN HEALTH
AND WELLBEING
A SYSTEMS APPROACH

中国科学院城市环境研究所
IUE
INSTITUTE OF URBAN ENVIRONMENT CAS

INTERNATIONAL
COUNCIL
FOR SCIENCE

iap
SCIENCE
RESEARCH
HEALTH
the interacademy partnership

UNITED NATIONS
UNIVERSITY

Foreword

With now over 50% of the global population in urban areas and this percentage projected to continue to increase, a research focus on urban areas and the health and wellbeing of their huge populations is central to the Mission of the International Council for Science which is "to strengthen international science for the benefit of society" and we mean all societies, including those urban societies around the world. One of our key priorities is Science for Policy and as part of that role, the International Council for Science has been the lead for the Science and Technological Major Group in many United Nations processes, including the negotiations of the Sustainable Development Goals. Many of the resulting SDGs relate to urban health and specifically Goal 11 on cities and human settlements and Goal 3 on healthy lives. The Council played a similar role in the negotiations of the Sendai Framework for Disaster Risk Reduction 2015–2030 which has as 2 of its 4 priority areas relating to understanding risk and risk governance. One of the Sendai Targets for Disaster Risk Reduction 2015–2030 is to lower average global mortality.

Recognizing that urban areas are often most vulnerable to hazards and where most of the people are, there is a clean connection with urban health and wellbeing. At the 21st Conference of the Parties under the UN Framework Convention on Climate Change in Paris, the Paris Agreement was concluded which includes Article 7, paragraph 1. "...enhancing adaptive capacity, strengthening resilience and reducing vulnerability ...". Article 5 states that: "... adaptation action... the best available science ... integrating adaptation into relevant socioeconomic and environmental policies and actions." Again, the urban health and wellbeing connections are clear.

These are complex intersecting issues—urban area, healthy lives, governance, disaster risk, resilient and reduced vulnerability, climate change adaptation, vulnerable groups, communities and ecosystems, indigenous people and local knowledge—and addressing them in effective ways to provide information and advice for policy makers is a challenge and also part of the Council's priority on science for policy. The program "Urban Health and Wellbeing: A Systems Approach" will bring together a research focus on these complex intersecting issues. Recognizing the human needs and challenges requires a systems approach that is transformative, transdisciplinary and integrative. Having the United Nations University and the InterAcademy Medical Panel as co-sponsored makes for a very valued partnership.

Another of the Council's priorities is International Research Collaboration. This has been a focus for many decades of sponsoring or co-sponsoring global research programs on the broad aspects of human-environmental interactions. As the global communities need to work together to address the intersecting issues of sustainable development, climate change and disaster risk reduction, the International Council for Science, as the common co-sponsor of each of the Urban Health and Wellbeing, World Climate Research Programme, Future Earth: Research for Global Sustainability and Integrated Research on Disaster Risk, will work with our programs and the global science community to deliver the information and outputs that will enable governments and societies around the world to effectively address these global challenges.

On behalf of the International Council for Science and its members, I thank the scientific leaders of the Urban Health and Wellbeing program for their insights and energies in bringing this program forward towards implementation and look forward confidently to seeing the scientific advances for the benefit of all.

Gordon McBean
President, International Council for Science (ICSU)

Preface

Cities are now humans' dominant habitat and urbanization is currently one of the most profound global development forces. As urbanization trends are projected to continue, reaching the global Sustainable Development Goals (SDGs) requires cities to be sustainable by ensuring that they are healthy places to live and providing opportunities to improve the wellbeing of its inhabitants. Human health and well-being in cities is closely linked to the structures and functions of cities. They provide a wide range of benefits and are supported by the global ecosystem. Despite being complex systems which provide goods and services, cities also have ecological footprints many times larger than the area they physically occupy. Therefore, for cities to become sustainable it requires to not only strive for better human lives but to see the city system embedded in the global ecosystem and to find a path from urban to planetary health. Urban and planetary health are inseparably linked. This is strongly emphasized in the many articles of the New Urban Agenda agreed on in Quito, Ecuador October 17–20, 2016, which refer to urban–rural linkages.

The definition and concept of health applied in this book can be seen as a natural integrator cutting across various urban sectors and relevant, as progress indicator, for many urban policies. Applying health as a natural integrator should be seen as an attempt to change from a siloed urban planning and management approach to an integrated one. In that context, the World Health Organization (WHO) paper entitled "Health as the pulse of the new urban agenda" includes contributions from members of this Urban Health and Wellbeing: a Systems Approach program. It clarifies the linkages between health and urban development and provides a vision for integrating health into urban planning and governance. The New Urban Agenda recognizes that urban development and governance can mitigate the risks and promote the health and wellbeing of urban populations. Implementing it requires agreement on definitions, concepts, contexts, frameworks and possible pathways of change towards urban health and wellbeing. This volume provides that foundation for implementation.

The systems approach we develop in this volume goes beyond the mere recognition that cities function as complex systems and are themselves part of the planetary ecosystems. It is rooted in complex system science, recognizes the knowable and unknowable, plannable and unplannable aspects of cities, suggests scientific and societal responses to both and the type of decision making required to improve the health and wellbeing of people in urban environments. That systems approach embraces ecological, economic and social determinants of health and offers a transdisciplinary approach for translating science into action. Without a systems approach to urban health and wellbeing, the sustainable urban development goals of the New Urban Agenda are unlikely to be achieved.

This book builds on the science plan (ICSU 2011), which led to the establishment of the program and conceptualises core elements for its implementation. It provides an overview of historical and current urban developments and related health challenges and proposes a theoretical and conceptual model for understanding how human health and wellbeing can be achieved in cities. Eventually it proposed guiding principles and thematic areas for a range of implementation activities. Implementing activities which enhance health and wellbeing in the changing urban environment by taking a systems approach requires clarity on what a system approach encompasses. This volume attempts to do so by defining a systems approach for urban health and wellbeing and by developing an innovative conceptual model which explains how cities function as complex urban systems for the health and wellbeing of its residents.

The vision of the program are cities functioning as integrated complex systems, which sustainably provide benefits for the health and wellbeing of its residents. The Urban Health and Wellbeing: a Systems Approach programme aims at promoting and coordinating research projects which take a systems approach to urban health and wellbeing, developing methodologies and identifying data needs, building and strengthening capacity for applying systems methods and thinking and communicating new knowledge on systems approaches to urban constituency groups. To become effective, new collaborative networks need to be established, which collaborate with ICSU member organizations and research programs, with universities, community initiatives, scientific projects, and educational as well as cultural projects. The program aims at facilitating the creation of networks and forums to promote the exchange of different types of knowledge in order to advance the evidence base for policy and decision making showing the added value of a systems approach for urban health and wellbeing.

Acknowledgements

This report is the collaborative outcome of the members of and advisors to the scientific committee and co-sponsors of the ICSU program on "Health and Wellbeing in the Changing Urban Environment: a Systems Analysis Approach": Indira Nath, Ana V. Diez Roux, Anthony Capon, Carlos Dora, Christl Donnelly, Gérard Salem, Hany M Ayad, Ilene Speizer, Jo Ivey Boufford, Keisuke Hanaki, Luuk Rietveld, Pierre Ritchie, Saroj Jayasinghe, Susan Parnell, and Yongguan Zhu. The book builds on inspiring discussions and exchanges with Trevor Hancock, José Gabriel Siri, Uta Dietrich, Barry Newell, Katrina Proust, Roderick Lawrence, Peter Head, Fabien Pfaender, Stefan Reis, Feng Feng, Yi Zhang, Armin Bobsien, Christoph Graf von Waldersee, Denise Young, Muhammad H. Zamman, Deyu Zhou, Basile Chaix, Vittoria Colizza, Martin O'Connor, Gabriele Harrer-Puchner and all participants of the science-policy dialogue on "Modelling Urban Health and Wellbeing" which took place on the island of Gulangyu, Xiamen April 28–29, 2016. Contributions from Lucilla Spini, Eulalie Ruan and David Marc Jones are highly appreciated.

This book has further greatly benefited from discussions with colleagues at ICSU's headquarters in Paris, ICSU's Future Earth program, ICSU's Regional Office for Asia-Pacific, the United Nations University International Institute of Global Health (UNU-IIGH) in Kuala Lumpur, the United Nations University Institute for Environment and Human Security (UNU-EHS) in Bonn, the United Nations University Institute for the Advanced Study of Sustainability (UNU-IAS) in Tokyo, the Center for Development Research (ZEF) at the University of Bonn, the School of Public Affairs at Zhejiang University in Hangzhou, the Complex City Lab at Shanghai University, and from the insights gained from discussions with the World Social Science fellows, who held a seminar on "Big Data in the Urban Context", hosted and co-organized by the Urban Health and Wellbeing program in Xiamen, December 2015.

As noted in the background chapter, hundreds of colleagues contributed to the development of the Urban Health and Wellbeing (UHWB) Program. Marvalee Wake and Thomas Roswall are identified in the background chapter. Carthage

Smith, ICSU Deputy Executive Director from 2002 to 2014, provided high-level substantive and technical support to the UHWB initiative throughout his tenure. José Siri‘s work as Research Associate in the several years that spanned the Planning Group and establishment of the Scientific Committee (SC) was appreciated by both entities. Dov Jaron co-chaired the 2008 Vienna workshop that demonstrated the scope of support and the potential for an interdisciplinary program dedicated to urban health and wellbeing using a systems approach. Gérard Salem was instrumental in organizing the 2013 Paris conference that confirmed the Program's potential at a critical moment and continues on the SC. Jo Ivey Boufford played a key role in facilitating the partnership with the InterAcademy Medical Panel, now the Inter Academy Partnership (IAP); Anthony Capon has played a similar role with the United Nations University as well as participating at earlier stages. Both are now members of the SC. Pierre Ritchie has contributed to each of the bodies created to further the UHWB initiative, from the initial meeting of Unions in 2002, to current service on the SC. Indira Nath participated at multiple stages and has provided essential direction as the initial Chair of the Scientific Committee.

We are particularly thankful for the igniting thoughts presented by all the participants of the first Xiamen Expert Workshop and inauguration ceremony of the International Programme Office in Xiamen, China, Oct 2014. We acknowledge the generous support of the Institute of Urban Environment (IUE), the Chinese Academy of Sciences, the Chinese Academy of Science and Technology and the municipality of Xiamen, without which the publication of this book would not have been possible.

Contents

Introduction to ICSU, UNU, IAP and IUE

International Council of Science (ICSU)

Founded in 1931, the International Council for Science (ICSU) is a non-governmental organization representing a global membership that includes both national scientific bodies (120 National Members representing 140 countries) and International Scientific Unions (30 Members). The ICSU 'family' also includes upwards of 20 Interdisciplinary Bodies—international scientific networks established to address specific areas of investigation. Through this international network, ICSU coordinates interdisciplinary research to address major issues of relevance to both science and society. In addition, the Council actively advocates for freedom and responsibility in the conduct of science, promotes equitable access to scientific data and information, and facilitates science education and capacity building (www.icsu.org).

UNU

The United Nations University (UNU) is a global think tank and postgraduate teaching organization headquartered in Japan. The mission of the UN University is to contribute, through collaborative research and education, to efforts to resolve the pressing global problems of human survival, development and welfare that are the concern of the United Nations, its Peoples and Member States. In carrying out this mission, the UN University works with leading universities and research institutes in UN Member States, functioning as a bridge between the international academic community and the United Nations system. Through postgraduate teaching activities, UNU contributes to capacity building, particularly in developing countries (www.unu.edu).

IAP

The InterAcademy Partnership (IAP), was formally launched in South Africa in March 2016. It brings together established networks of academies of science, medicine and engineering, namely IAP, the global network of science academies, the InterAcademy Medical Panel (IAMP) and the InterAcademy Council (IAC). Under the new InterAcademy Partnership, more than 130 national and regional member/academies will work together to support the special role of science and its efforts to seek solutions to address the world's most challenging problems. In particular, the new IAP will harness the expertise of the world's scientific, medical and engineering leaders to advance sound policies, promote excellence in science education, improve public health, and achieve other critical development goals.

IUE

The Institute of Urban Environment (IUE) of the Chinese Academy of Sciences (CAS) was established on 4 July 2006. IUE-CAS is a national research institute engaged in comprehensive studies on the world’s urban environment. IUE is a multidisciplinary research institute engaged in fundamental and applied research on the cutting edge of environmental and resource sciences, as well as the development of new technologies for environmental remediation and waste management, which encompasses the impacts of urbanization on ecosystems and the social economy.

Advancing Health and Wellbeing in the Changing Urban Environment

1 Background

The genesis of an interdisciplinary program on urban health and wellbeing was a meeting in Rio de Janeiro, Brazil on the occasion of the 27th General Assembly of the International Council for Science (ICSU) in September, 2002. With the support of ICSU Executive Director, Dr. Thomas Roswall, the meeting was convened by Professor Marvalee Wake, President of the International Union of Biological Sciences. It was attended by representatives of nine ICSU member Unions. This is the first recorded discussion within ICSU of health as an important focus for science.

As noted in the summary prepared by Professor Wake, this initial meeting struck a bold vision for health as an area of common legitimate interest for the international science community. The nine Unions agreed to initiate the development of an action plan anchored by the theme of 'Science for Health and Wellbeing'. From the outset, health and wellbeing were both regarded as the fundamental pillars for building a common vision and undertaking collaborative, interdisciplinary activities. A second outcome of the Rio meeting was a commitment "to facilitate interdisciplinary by forming effective partnerships that transcend disciplinary borders."

Only a few months later, a second meeting occurred at ICSU headquarters in Paris in February, 2003. Unlike the short meeting in Rio, this was a two-day meeting. It was attended by representatives of ten ICSU Unions and three ICSU affiliated interdisciplinary bodies. Although the meeting was structured around a detailed agenda, it was characterised as a 'brain-storming session to delineate the scope of the initiative and to formulate a mission statement, as well as potential areas for collaboration'. More than twenty focal topics were identified.

F.W. Gatzweiler et al., *Advancing Health and Wellbeing in the Changing Urban Environment*, Urban Health and Wellbeing, DOI 10.1007/978-981-10-3364-3_1

The more immediate contribution of the second meeting was agreement on a statement of scope and mission for the initiative:

> We recognise that the health sciences require the partnership of a broad range of sciences and technologies to advance human health and well-being. The initiative 'Science for Health and Well-being' is a collaborative effort by ICSU Unions and other ICSU bodies to promote the elaboration and utilisation of the scientific information appropriate to understanding human and environmental health and well-being in its many dimensions.

The goals of the initiative are to: 1. demonstrate the extent to which a range of science and technology is important to health and well-being; 2. educate (at multiple levels) about science and technology using modern means of communication; 3. Collaborate to identify areas unmet and produce new ideas, science and technology partnerships to look to the future; and, 4. develop an inventory of ongoing programmes, activities in order to identify unmet needs; in order to empower the science community, the public and policy-makers and in all countries with the knowledge base to affect their health and well-being.

While focal topics varied to some degree over time, it is remarkable that the broad vision was sustained over the dozen years between the initial meetings organized by the initiating ICSU Unions and the official launch of the UHWB International Program Office in 2014. In particular, a broad and encompassing understanding of health, a commitment to partnerships, and using pertinent scientific information to empower the public and policy makers have been embedded in all the work that led to the creation of the Programme.

A full review and analysis of the steps taken between 2002–03 and 2014 will not be offered here. Nonetheless, the main steps of what is best construed as an iterative process merits brief mention.

Following the 2003 meeting, the next major event was convened in conjunction with the 2004 ICSU mid-triennium Unions meeting. Hosted at the French Academy of Medicine, representatives from eighteen ICSU Unions and three ICSU interdisciplinary programs participated. The increasing attendance reflected a growing interest in the ICSU community in health as potential area for common work. ICSU Executive Director Thomas Roswall stated that the Health and Wellbeing initiative was consistent with ICSU's efforts to strengthen its collaboration with scientific fields beyond those in which ICSU had traditionally concentrated its efforts. He noted achievements in other areas, but observed that these had yet to include the health sector.

The following year, 2005, was a critical one in significantly moving the initiative forward. ICSU's Committee on Scientific Planning and Review (CSPR) approved a grant submitted on behalf of multiple Unions to support further work aimed at laying the groundwork for a comprehensive science driven health and wellbeing program. The letter announcing the grant's approval accurately reflected the internal debate at ICSU about making health a strategic priority. It noted 'the enthusiasm of CSPR for an initiative that builds on the strengths of so many Unions', as well as the committee's words of caution that 'there is a long way to go before this can be considered as a cutting edge ICSU programme'.

CSPR was concurrently working on ICSU's first ever Strategic Plan. Later in 2005, at the 28th ICSU General Assembly in Suzhou, China, a Strategic Plan (2006–11) was adopted. It included explicit provisions addressed to health, with an overall goal "to ensure that health considerations are duly taken into account in the planning and execution of future activities by building on the relevant strengths of Scientific Unions and Interdisciplinary Bodies". The associated specific action was the establishment of a Scoping Group to more clearly define how ICSU might contribute to science for human health. The Scoping Group had a dual role: (1) to liaise with and ensure coordination, where necessary, of the major health initiatives that were already being developed within the ICSU community; and (2) to identify additional areas or approaches where ICSU might add value to these initiatives.

While there was debate about whether health should be a scientific focus for ICSU, the outcomes of the General Assembly's deliberations confirmed that the initiative was no longer anchored in a sub-group of ICSU's Unions. There was now clear interest among a growing number of Unions as well as widespread support from ICSU's National members.

The Scoping Group worked during the 2006–07 period. Its final report was subsequently endorsed by CSPR and the ICSU Executive Board. It set the stage for shifting the focus from 'whether' ICSU should develop a health and wellbeing program to 'what' such a program should be and 'how' it should function. Its conclusions and recommendations were clustered in two parts.

One set of recommendations set the criteria for what the program should be. The Scoping Group proposed that an ICSU health initiative should:

- be international (relevant to more than one region) and interdisciplinary;
- build on the synergies between existing activities and interests of the ICSU membership;
- add value to other ongoing or planned activities;
- fulfil a unique niche that other international initiatives do not address;
- incorporate links with pertinent international bodies and/or networks outside the ICSU family, as necessary;
- generate new scientific insights of significant relevance to health (and other) policies, with the policy audience(s) clearly defined at the outset;
- focus primarily on population health/disease prevention and promotion of health and wellbeing rather than medical treatment and cures;
- fit within the context of the Millennium Development Goals and the broader agenda of science for sustainable development.

Once again, caution was urged about the pace of moving forward:

Given that ICSU does not have an established track-record or 'name' in the international health arena, it was also recognized that the rapid development of a major new research programme was unlikely and that a more cautious step-wise, iterative, approach was more realistic.

The second set of recommendations addressed the particular mission of an ICSU program on health wellbeing and its methodological orientation. The Scoping Group proposed that cities and the changing urban environment be the specific focus of the program. The compelling global context for this choice was the rapid growth of cities that would only likely accelerate further in the next quarter century. Embedded in ICSU's core value of responsibility to society, the compelling science rationale was the potential for interdisciplinary scientific contributions to help policy-makers make better sense of their communities in order to usefully inform their future decisions related to the health and wellbeing of people living in urban environments.

The proposed methodological orientation was a systems analysis approach to human health and wellbeing. No aspect of the work that contributed to the establishment of the Urban Health and Wellbeing Program has received greater attention, been more discussed and debated, experienced more misunderstanding, nor undergone such sustained evolution as has the systems analysis dimension. It may well continue to do so. A future historian of science who takes the time to trace the discussions and debate found in the primary archival sources on this matter will be well rewarded. For the moment, it is sufficient to note that systems analysis remains an integral component of the program, albeit with a current definition and scope that provides greater opportunity to adapt the particular application of systems analysis to a broad range of challenges, problems and questions.

On the strength of the Strategic Plan adopted for the period 2006–11, ICSU's Executive Board accepted the Scoping Group's recommendation to proceed with a full-scale planning exercise to design and propose the framework for a new ICSU interdisciplinary program on 'Health and Wellbeing in a Changing Urban Environment' using a systems analysis approach. In April, 2007, it mandated CPSR to establish the Planning Group. The Planning Group worked from January, 2008 to May, 2010. Its initial meeting was held in Vienna, Austria in conjunction with a workshop co-sponsored by ICSU and the International Institute for Applied Systems Analysis. The workshop was titled 'Towards a Systems Analysis Approach to Health and Well-being in the Changing Urban Environment'. Some forty participants from nineteen countries were drawn from the ICSU and IIASA communities. In particular, it abundantly re-confirmed the engagement of both ICSU's Unions and its National Members.

The Planning Group held five additional meetings. It submitted an interim progress report to the 29th ICSU General Assembly in Maputo, Mozambique in 2008 and to several ICSU Union meetings over the period it existed. Two of the Planning Group meetings included dedicated contiguous meetings with urban officials in France and China. These occasions provided an opportunity to better understand the kind of science-driven information that could assist policy makers in making more informed decisions that affected health and wellbeing. The Planning Group re-affirmed that the new program should generate policy-relevant knowledge that will improve health status, reduce health inequalities and enhance the wellbeing of urban populations. It foresaw a ten-year initial commitment to establish and implement the program.

The primary outcome of the Planning Group was the Science Plan that would serve as the substantive framework for the program. It was the core document submitted to the 30th ICSU General Assembly held in Rome, Italy in September 2011. Notwithstanding the earlier debates in the ICSU community and the admonition to proceed cautiously, the ICSU General Assembly overwhelmingly approved the establishment of the new program. On the basis of the proposed Science Plan and the provisions of the second ICSU Strategic Plan (2012–2017), the General Assembly endorsed the CSPR and Executive Board recommendations to establish a new interdisciplinary research initiative of ten-years duration on Urban Health and Wellbeing with a systems analysis approach, in collaboration with other partners; and to recognise the initiative as an Interdisciplinary Body. It also approved using ICSU's regional structures to ensure the involvement of developing countries. Since the original Science Plan developed by the Planning Group serves as the nucleus for the implementation activities presented in this document, it is not further elaborated in this background chapter.

Once the ICSU General Assembly approved the establishment of the Urban Health and Wellbeing program (UHWB), four essential steps remained with a fifth regarded as optimal. The first step was to set the Terms of Reference for the UHWB's Scientific Committee (SC) and the appointment of its initial members. The SC's Terms of Reference grant it responsibility for the scientific planning, coordination, guidance and oversight of the Programme, as well as monitoring the development of the programme over time, identifying gaps and priorities as necessary. Its specific mandates are (1) to integrate global and regional programme activities under a single coherent framework; (2) engage with relevant intergovernmental and non-governmental partners to promote the programme and shape its directions; (3) facilitate the involvement of interested ICSU Union and National Members and other Interdisciplinary Bodies; (4) play an advocacy role with relevant research public and private funders and donors.

In performing these functions, the Scientific Committee (SC) is supported by the International Programme Office. While the Terms of Reference were quickly developed by CPSR and ratified by ICSU's Executive Board, the second step of appointing initial SC members took somewhat longer than originally anticipated. Consequently, the first meeting of the SC took place in Paris only in December 2012. Thereafter, the pace quickened and key activities happened more expeditiously. A second SC meeting in Kuala Lumpur, Malaysia occurred in May 2013 in association with related events organized by the ICSU Asia-Pacific Regional Office.

The third SC meeting in September 2013 was contiguous with a large international conference on 'Urban Dynamics and Health'. Under the aegis of the new ICSU Urban Health and Wellbeing Program and hosted by the University of Paris-Ouest Nanterre La Défense and its Laboratory on Space, Health and Territories (LEST), this seminal event was co-sponsored by the French National Network of Local Decision-makers for Promoting Local Public Health (ESPT), and the French Institute for Research Development with contributions from ten international, European and national organizations as well as nineteen universities and research institutes from Africa, Asia-Pacific, Europe and North America.

The SC has held five additional meetings (typically two per calendar year) through November 2015. Working from the Science Plan produced by the Planning Group, the constant element in the SC's endeavours has been the creation of an action-oriented science and related policy agenda for a systems approach to urban health and wellbeing. Its work is partly reflected in the activities that completed the essential steps and production of this book. Related accomplishments and future outcomes are also described here.

The third essential step was selection of the country and venue for the International Program Office (IPO). The initial Call was issued by ICSU in 2013. The vetting of applications as well as related site visit was conducted by a special committee composed of persons drawn from CSPR, the SC and ICSU senior management. The process was concluded in late 2013 with the Institute of Urban Environment (IUE) of the Chinese Academy of Sciences (CAS) located in Xiamen, China selected as the venue for the International Programme Office for UHWB.

Once the venue for the IPO was determined, it was possible to initiate the final essential step, the recruitment of an Executive Director (ED). Under the authority of the SC, the ED provides sustained scientific and operational leadership for the Program. Dr. Franz Gatzweiler was chosen as the UHWB Program's Executive Director and he took up the post in October 2014.

Only two months after the ED arrived in Xiamen, the IPO was officially launched in December 2014 highlighted by the Xiamen Expert Workshop on Systems Approaches to Urban Health and Wellbeing. The formal opening of the IPO, inaugurated by ICSU President Gordon McBean and CAS Vice-President Jinghai Li, was attended by other local and national officials as well as the participants of the Experts Workshop. More than fifty participants from all regions of the world covering a comprehensive range of expertise pertinent to urban health and wellbeing as well as systems approaches engaged in lively exchanges through 22 ignition talks, 5 plenary sessions, 4 breakout work groups and one plenary workgroup on research priorities for the program.

The optimal step noted above was also achieved. One of the recommendations of the Planning Group, affirmed in the SC Terms of Reference, emphasized the importance of developing the program in association with partners. The United Nations University and the Interacademy Medical Panel have become affiliated as co-sponsors. They will contribute to the UHWB Program's overall evolution and future direction. Together with other partners, they will also contribute to focal research and capacity-building activities.

It took a dozen years to move from the small band of ICSU Union's bold vision for an international science program dedicated to health and wellbeing. Hundreds of colleagues contributed along the way. A few provided sustained commitment and leadership at virtually every stage. The next stage is now set to vigorously pursue the potential captured in the original Science Plan and now operationalised in this publication.

2 Introduction: Urban Transformations and Health

The World's population is growing and the world is becoming increasingly urban. In 2012, the world counted 7 billion people with about 54% living in cities. In 2050 world population will have reached 9.5 billion with an estimated 66% (6.3 billion) living in cities (United Nations 2014). This makes urbanisation a global mega-development which needs to be better understood in order to provide aspired levels of human health and wellbeing.

Living in urban areas generally comes with advantages for human health and wellbeing from population densities being higher, physical, technological and institutional infrastructure being denser, innovation and creativity being stimulated. Further, public service provision and employment opportunities tend to be better in cities and economic growth is higher. Despite the fact that "... human health is better now than at any time in history ..." (Whitmee et al. 2015), there is a growing recognition that this progress has come at social and environmental costs, such as increasing inequality (Unitd Nations 2013), increasing energy use and related greenhouse gas emissions, air, light and noise pollution, soil degradation, biodiversity loss, and severe water stress. Within cities there are also large variations in health and wellbeing. Those challenges are a threat to urban health and wellbeing and the "urban advantage"—the health benefits of living in urban as opposed to rural areas—has to be continuously created and maintained (Rydin et al. 2012) as urban environments change. Therefore, it is paramount to better understand the complexities of changing urban environments and their impact on health and wellbeing.

Generally, cities tend to become flatter (sprawl), greyer (ageing) and poorer. A one per cent decline in densities per annum between 2000 and 2050 increases the urban land area by a factor of four[1]. Based on satellite imagery, Angel et al. (2005) found that cities around the globe are sprawling outwards. Globally, population densities in cities are declining. Despite variations and exceptions, the overall trend is that cities are becoming flatter, i.e. growing outwards towards the periphery. As infrastructure coverage tends to decrease towards the periphery, dependence and use of cars increases and access to public services decreases.

In 2002, about 750 million urban dwellers of low-income countries lived below the poverty line of $2/day, which represented about one third of all urban residents. Almost half of the world's urban poor were living in South Asia and another third in Sub-Saharan Africa. From the 1990s to 2000, the pace in urban poverty reduction has been slower than reductions in rural poverty indicating an overall urbanisation of poverty. Although poverty has generally declined in rural and urban areas the share of the total poor living in urban areas has been rising in all regions within the same time period (Baker 2008; Caldwell and Caldwell 2002). In 2011 one third of all urban residents (828 million people) lived in slum conditions and this proportion is expected to remain constant, with urbanisation the number of people living in poverty in cities will rise. The urban poor, especially children under five years, are exposed to health risks and carry a "double burden of disease" by being exposed to an increasing prevalence of infectious and chronic diseases as well as exposure to violence and unsafe living conditions (Coast 2011).

Since the 1970s there has been a fertility decline in all regions of the world and populations around the world are rapidly ageing. The number of people aged 60 years and above will reach nearly 2 billion, with low-middle income countries experiencing the most dramatic increase in the proportion of older people, which will rise to approximately 80 percent (WHO 2015; Beard and Petitot 2011).

With people in cities ageing, cities flattening (sprawl) and becoming poorer, cities and people need to transform for maintaining and improving health and wellbeing. The emerging health risks include road traffic injuries and deaths, urban violence, obesity and other non-communicable diseases, new or re-emerging infectious diseases and health risks from living in unsafe settlements.

However, as global and urban transformations represent risks to human health and wellbeing, cities are also the places most likely to generate the solutions. The city is the most likely place in which connectedness is enhanced, where human brains come together and compete, and where innovation and creativity emerges. Ideas, inventions and innovations emerge in cities and can be part of the solutions required for the urban health challenges (Bettencourt 2013).

We are, however, confronted with yet another global transformation in science and research. Given the dynamic and complex nature of global change processes, the process and organization of how we achieve and communicate knowledge, do research and create collective intelligence is undergoing radical changes[2]. Uncertainty is at the edge of knowledge and cutting-edge science is science which addresses uncertainty. Science is evolving from the search and production of precise, proven and indisputable results by an academic elite to a collective effort of producing knowledge, in an interdisciplinary manner[3], which is able to better respond to the global mega-challenges of our time. Not only different types of knowledge engage in this process of co-production, also processes and methods for organizing the collective production of knowledge are changing.

Conversations from local to global scales are taking place about the right way to live in our new urban environment and with our fellow humans. The UN Habitat[4] gatherings and the collective design of the post-2015 UN Sustainable Development Goals (SDGs) are important global efforts being made. The targets of the SDGs and in particular Goal no. 11 on "Sustainable Cities and Communities"[5] all refer to services which also urban systems should be providing for health and wellbeing: housing, transport, culture, recreation in green public spaces, avoidance of pollution, especially air and water, and inclusive planning. All of these are also important physical, social and mental determinants of health and they are part of the conceptual framework we have developed for this programme.

ICSU's programmes aim at contributing to universal use and equitable access to scientific data, data services, products and information[6] and at improving the quality, reliability, management and accessibility of data to all fields of science, technology, society and the environment[7]. This programme is an interdisciplinary, global science programme which brings emerging issues of urban change and health into public discourse and onto political agendas by building collective intelligence, through research, networking, engagement and communication between stakeholders in research and education, civil society, and professionals in order to sustain the health of people and the planet in an urban world.

Purpose and Target Audience
This book aims at providing a background, rationale and conceptual framework for a systems approach to urban health and wellbeing. This plan aims at presenting urban health and wellbeing as a challenge of understanding and governing cities as complex urban systems. As such they require governance responses which can harness and make best use of the complexities of complex urban systems.

The target audience of this publication are scientists and experts, educators and teachers, students as well as decision makers from policy and civil society, including the media, governmental and non-governmental organizations and urban communities who aim at improving the health and lives of people in cities. Diverse urban stakeholder groups may use this report as a guide for discussion and deliberation to co-create the cities they want to live in. As it provides a theoretical and conceptual framework which can be operationalized, the book can be used as a reference for research activities, as a guide for policy making as well as teaching graduate and undergraduate university courses in higher education institutions.

The ICSU science programme on urban health and wellbeing offers an opportunity to reflect on the types of problems challenging health and wellbeing in complex urban systems and the type of knowledge, which is required to address them. Further, it offers a conceptual framework that structures the multifaceted and multi-disciplinary issues of urban health and wellbeing. Having a conceptual and theoretical model is essential for implementation. By perceiving urban health issues such as air pollution, energy and transportation networks as interconnected elements of complex urban systems, science has the opportunity to better understand the systemic role urban (sub)system components play and their impact on health and wellbeing. Based on such an understanding urban policies tend to cause less costly side-effects for human health. Communicating and co-producing this knowledge can lead to better governance of cities towards sustainable development.

3 Context and Relevance: Why Cities and Why Urban Health?

3.1 Urbanisation, Urban Density and Complexity

Urbanisation is currently one of the earth's most intense development challenges, with drastic implications and real pressures on socio-ecological sustainability, and the health and wellbeing of a major proportion of global population. More than half of the global population, about 3.5 billion people, live in cities. Urban life contributes to 80% of global GDP but also to 70% of anthropogenic carbon dioxide emissions globally every year[8]. Cities pose enormous challenges to planetary health as well as human health and wellbeing which extend to global-scale phenomena such as air and water pollution, climate change and biodiversity loss. Those developments compromise the capacity of ecosystems to deliver the supporting goods and services (Alberti 2010) for present and future generations' living in cities.

Global sustainable development challenges concentrate in cities and are closely tied to human health and wellbeing. For that reason, two Sustainable Development Goals[9] address those concerns: Goal 3: "Ensure healthy lives and promote well-being for all at all ages" and Goal 11: "Make cities inclusive, safe, resilient and sustainable". The health impacts of urban development (Vlahov et al. 2007), however, need to be acknowledged and addressed by a comprehensive systems approach—which is the intention of this programme.

Many cities are among the most vulnerable geographical sites threatened by the effects of global warming through sea level rise, temperature changes, changes in biodiversity, extreme and uncertain weather conditions, diseases and social conflict. Globally, cities are nodes of multiple networks (hotspots) where flows of resources, information, power and energy come together, condense, accumulate and are transformed and dissipate (Image 1). Similarly, infectious pathogens come together in cities, transmit, and then spread elsewhere.

Human health and wellbeing in cities is under pressure from the need to accommodate increasing numbers of people, growing numbers of people moving to cities, including climate refugees (Biermann and Boas 2010). As access to health care and employment opportunities and other public services is generally better in cities than in rural areas, cities will need to adjust their infrastructures and processes by strengthening adaptive capacities (Satterthwaite et al. 2009) through better understanding cities as complex systems (Gurr and Walloth 2014; Albeverio et al. 2007). Cities are places from which societies are governed and which are governed

Image 1 Cities are hotspots of activity and energy, which makes them visible from space as illuminated spots at night on earth[10]

by societies, and where social, cultural and economic development takes place in cycles of progress and decline, beyond their administrative or physical boundaries.

In 2008, for the first time in history urban populations overtook rural populations (Fig. 1). Cities have become the dominant human habitat. Their number and size is expanding in all regions of the world with China as the world's largest urban nation. Whereas from 1990–2014 growth rates of urban agglomerations were highest in Asia, India and western Sub-Saharan Africa (3–5%), urban growth rates are slowing generally but are relatively high (1–5%) in Sub-Saharan Africa.

The crux and a major challenge of urbanisation is that it poses risks and opportunities to health and wellbeing, which triggers the need for better understanding the dynamics of urban systems; their density, connectedness, and flows of people, resources, data and energy.

Cities are complex urban socio-ecological-technological systems providing multiple functions for human health and wellbeing. The organisational opportunities and challenges in cities increase with size, scale and complexity. In doing so they can produce negative effects (externalities) which harm human health and wellbeing and the global ecosystem in which they are embedded. There is much uncertainty about what type of complex system cities are and how they behave. In her landmark book on the "Life and Death of Great American Cities", Jane Jacobs (1961: 433) explains that cities need to be viewed as organized complexity:

> Cities happen to be problems in organized complexity, like the life sciences. They present "situations in which a half-dozen or even several dozen quantities are all varying simultaneously and in subtle interconnected ways." Cities, again like the life sciences, do not exhibit one problem in organized complexity, which if understood explains all. They can be analyzed into many such problems or segments which, as in the case of the life sciences, are also related with one another. The variables are many, but they are not helter-skelter; they are "interrelated into an organic whole." ... Consider again, as an illustration, the problem of a city neighbourhood park. .... No matter what you try to do to it, a city park behaves like a problem in organized complexity, and that is what it is.

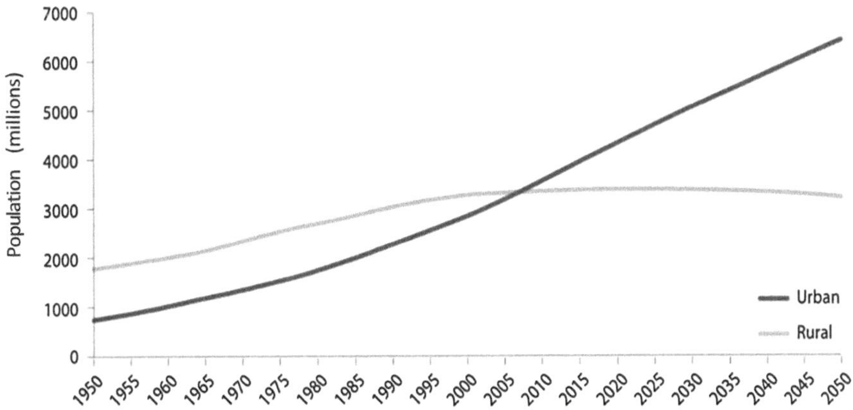

Fig. 1 Urban and rural population in the world 1950–2050 (United Nations 2014)

Cities are not only complex but also dynamic systems.[11] The functions they perform deliver valued services at one point in time but can have adverse effects in the future. Urban economic growth and productivity can be an attractor for employment but also increasingly unattractive when people become aware of the negative ecological and health impacts.

With rapid urbanisation, failures of understanding and managing urban complexity for human health and wellbeing can lead to foregone opportunities and increasing risks for human lives in cities. Historically, ancient (low density) megacities[12] have declined rapidly as their ecological footprints became too large or their responsiveness to climate and environmental change too slow. Their rulers and inhabitants failed to recognize ecological relationships between forests which were cut down for energy supply and changing micro-climates which led to droughts and harvest failures[13]. Today we have the means and technology to better understand the complex relationships and embeddedness of urban systems into global ecosystems which sustain them.

Increasing density, connectivity and complexity (Batty and Marshall 2012), when viewed as a challenge of complex systems, improving health and wellbeing in cities can be achieved by better governance of urban complexity. Improving the city's surveillance and response capacities—not only for diseases in healthcare systems but also for social, economic and ecological determinants of health, is one way to harness urban complexity.

As urbanisation has negative and positive impacts on health and wellbeing, it becomes important to understand the dynamics of urbanisation from a complex systems perspective. A key causal factor appears to be connectivity. Better connectivity does not only refer to better and denser transportation infrastructure. Connectivity also plays a role in communication, data sharing and institutional infrastructures. The links between physical and social infrastructures seem to play an important role for making best use of urban potentials. High connectivity creates opportunities because people benefit from the availability, access and use of urban system services and flows of resources (Andris and Bettencourt 2014). High connectivity paired with inclusion makes people part of the city system and then become agents of urban change. It allows people to make use of their capabilities, communicate, understand what is happening and shape the urban system according to what they value being and doing.

UN Habitat estimates that the number of urban residents living in slums in Africa will increase from 924 million in 2001 to 2 billion in 2030[14]. Changing socio-economic conditions in cities such as increasing inequality are mirrored by profound health inequalities (Harpham and Tanner 1995). Improved access to housing and health care for the rich does not solve the problems of increasing homelessness and infectious diseases (Harris and Keil 2008) among the poor.

Cities are associated with a substantial part of global energy consumption and energy related greenhouse gas emissions and fewer than 500 cities will account for over 60 percent of global income growth and 50 per cent of energy-related greenhouse gas emissions growth between now (2014) and 2030[15]. Butler and Harley (2010) identify three broad causal pathways by which changes in climate will impact on urban health:

1. Direct health impacts and threats from extreme weather events, heatwaves, worsened air pollution and sea level rise for coastal cities, injuries, death, and heat stress.
2. Indirect health impacts from changes in biological and ecological processes, for example, food and water quality, infectious diseases and malnutrition.
3. Health influences from social, infrastructural and economic conditions, causing mental stress, trauma and conflict.

Those cause-effect chains are indeed better understood as positive and negative feedback loops. For example, Brown et al. (2011) have mapped the positive and negative feedback loops of climate adaptation measures for human health in an urban setting. Figure 2 shows three reinforcing (positive feedback) loops (R1, R2, R3), and one balancing (negative feedback) loop (B). These loops are often important drivers of system behaviour which makes them to possible levers of change. The heavy arrows, which form the positive feedback loop R1, represent an important dynamic of the climate system, whereby rising temperatures increase the fraction of the community suffering from thermal stress. As this fraction increases, the use of air-conditioning rises (all else being equal) (B). The use of air-conditioning increases the load on power stations, which contributes to an increase in the production of greenhouse gas (GHG) emissions and atmospheric GHG concentrations, which eventually leads to further increases in temperatures.

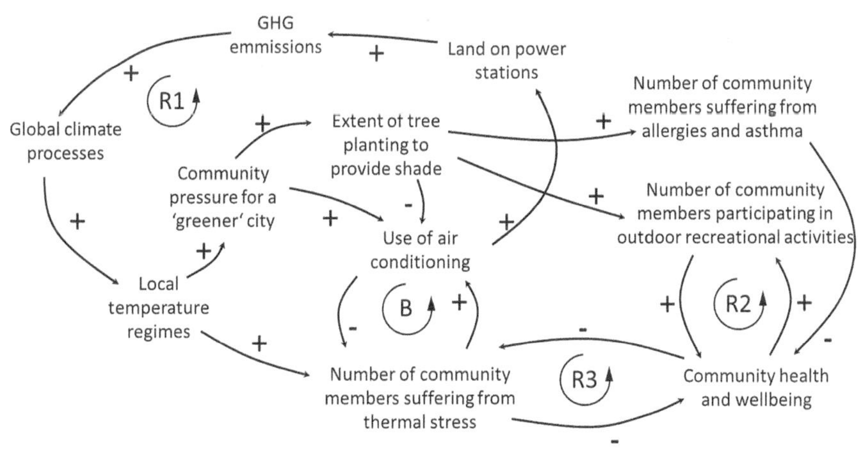

Fig. 2 Mapping the dynamic complexity of thermal stress, tree canopy in streets, climate change (Modified from Brown et al. 2011)

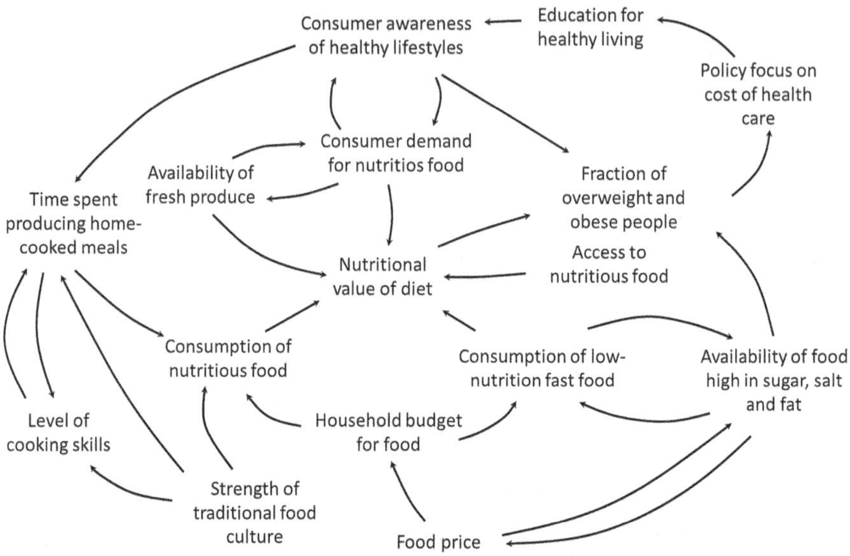

Fig. 3 Dynamic relationships between variables for food security and the proportion of obese people in urban communities (Modified from Proust and Newell 2016)

Figure 3 demonstrates how the nutritional value of a particular diet, consumer awareness of healthy lifestyles, the availability of food which is high in sugar, salt and fat directly influence the fraction of overweight and obese people in an urban community. It also shows how multiple indirect variables such as food prices, level of household cooking skills, traditional food cultures or time spent for preparing the food consumed are interconnected.

Understanding the dynamic complexity of urban systems which may produce unintended health outcomes can improve the quality of decision making by making the interconnectedness of system variables visible and showing options for interventions (Fong et al. 2009).

Urbanization and urban change are not the only factors, which influence the health of urban populations. In recognition of the global force that Homo sapiens has become, the epoch we are currently living in has been referred to as the Anthropocene. It is an epoch in which the amplified global circulation of bioactive nitrogen compounds, the loss of biodiversity, and human-induced climate change have reached levels that will leave traces in geological time spans and pose fundamental threats to human well-being and health (Rockstrom et al. 2009; McMichael and Butler 2011; McMichael 2013; Whitmee et al. 2015). It implies, that human health cannot be secured without achieving the healthy functioning of planet earth[16] (Image 2).

Image 2 City lights of Asia and Australia[17]

3.2 Governing Urban Complexity for Health and Wellbeing

Governing[18] urban complexity for health and wellbeing means governing dynamic interactions between people, ecology, and technology. Governance failures and failures to predict and prevent adverse health and wellbeing outcomes in complex urban systems can impose great costs on society[19] and have much to do with socio-political and economic pressures, power, and the ability and willingness to understand such problems as belonging into the realm of complexity. Governing dynamic complex urban systems for improving urban health and wellbeing therefore requires a better understanding of how multiple urban subsystems interact and how interventions and institutions inhibit or enable connectivity and interactions in various socio-cultural, political, or economic interaction domains and multidimensional spaces[20].

Governance styles are strongly determined on how the dynamics and temporality of change of a system in question are perceived and the respective styles of strategic action (Stirling 2006) (Fig. 4). System changes can be understood as short term shocks or longer term stresses. The styles of action can be control oriented or adaptive management styles (Weick and Sutcliffe 2001). Control oriented styles of

Fig. 4 Governance styles are determined by styles of action and how the temporality of system changes are perceived[21]

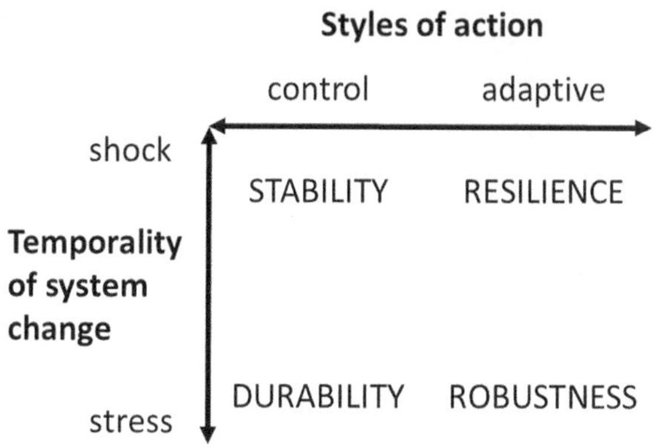

governance assume less uncertainty and more calculable risks. They assume that stability (against shocks) or endurance (against stresses) of a system need to be maintained or restored by control and order. Responsive or adaptive styles of governance acknowledge limits to control. Actions which aim at maintaining the resilience of a system build on the belief that the system changes are short-term shocks. Actions which aim at system robustness perceive changes as long-term stresses, for example, changes in the human immune system resulting from less human-animal contacts or less exposure to nature (Rook 2013).

In practice there is a tendency for styles of action to shift towards stability (Leach et al. 2010), for example by rapid response and containment of disease outbreak such as Ebola in Africa (Heymann et al. 1999). Further, creating stability is associated with routines and institutions (North 1990). Institutions of sustainability (Beckmann and Padmanabhan 2009) however, recognize that rules sometimes emerge in action situations characterised by value plurality, complexity and radical uncertainty—situations that cannot be planned, controlled or predicted. The World Health Organisation's Global Outbreak Alert and Response Network is an example of governance strategies for resilience. It is set up to respond to unpredictable external shocks (outbreaks) with a flexible response network that can be mobilised when needed.

Improving resilience of cities can also compromise the health and wellbeing of some segments of urban populations, especially when resilience is achieved by specialisation, diversification and some unequal parts of society (Béné et al. 2012a, b; Davidson 2010). Olsson et al. (2015) noted that "One person's resilience may be another person's vulnerability". The ways in which health and wellbeing in cities can be achieved and maintained, therefore depends on a systemic understanding of the issues at stake. Resilience is just one of many possible system states and behaviours.

Improving our understanding of how cities function as complex dynamic systems also means scaling up the ability to produce, monitor, predict, process and respond to changing information for an adaptive advantage. Intelligence has been defined by those abilities and it has been suggested that urban systems achieve some level of intelligence in order to solve complex problems as a collective entity

(Maahsen-Milan et al. 2013). The idea of an intelligent, sustainable city includes cities which are green and smart, which integrate core areas (such as energy, mobility, planning, governance and economy) and include cross-sectional topics (such as ICT, innovative finance and public participation)[22]. Technology is a critical tool for achieving the vision of an intelligent city[23]. "... assuming that greater populations will find themselves residing in networked, intelligent megalopolises, we need to give more serious consideration to designing urban interfaces for urban citizens, who have a right to know what's going on inside those black boxes—a right to engage with the operating system as more than mere reporters-of-potholes-and-power-outages. We need to focus attention on the "bleed points" between the concrete, digital and social city, those zones where citizens can investigate the entwinement of various infrastructures and publics." (Mattern 2014).

Finally, although health-oriented policy-makers have advocated for an approach that "government objectives are best achieved when all sectors include health and well-being as a key component of policy development" by taking a "Health in all Policies Approach"[24], there is no general consensus on the importance of health to urban growth and development and how specifically to govern diverse types of cities to achieve urban health and wellbeing outcomes, especially in light of uncertainty[25] (Funtowicz and Ravetz 1990).

Conventional urban planning strategies typically make predictions which are extrapolated from past trends and urban governance tend to control and order different styles of action. One way of responding to uncertainty is by identifying common myths (Alberti 2008), such as that urban systems retain stability of a steady state by means of optimisation strategies. Optimisation, however, relies on being able to quantify risks, which is not possible under radical uncertainty. Transferring planning strategies from one place or scale to another is problematic or impossible because changes in scale, space and time influence the urban systems' behaviour (Walker and Meyers 2004).

Research will need to shed more light on how and under which circumstances adaptive, inclusive urban governance for human health and wellbeing can be implemented. "In building institutions and strategies for resilience and robustness, insights from work on adaptive governance in the context of ecosystems are helpful, emphasising the value of self-organising and self-enforcing networks of individuals, organisations and agencies that have the capacity for flexible, collaborative and learning-based approaches." (Leach et al. 2010: 375).

4 Policy Relevance and Decision Making

Decision makers from local to national governments and civil society need to be able to decide and act in situations of complexity and uncertainty (Mehta et al. 1999; Uhl-Bien et al. 2007). Seeking ways to reduce uncertainty by generating knowledge and inferring rules for decision-making is one strategy to decision making, which corresponds to one part of this programme's "systems approach". It aims at avoiding ambiguity and defining behavioural rules which respond to the patterns, regularities and relationships identified in complex systems by science. Ambiguity is not only avoided because of ignorance but also because of the sunk cost effect: past investments in establishing the present "law and order" or status quo lead to avoidance of change (Janssen and Scheffer 2004). Those strategies correspond to governance strategies which lean towards stability. They aim at pushing the frontier of knowledge as far as possible and increasing the set of information and knowledge towards what is possible needs to be revealed.

Another strategy is accepting uncertainty and responding with emergent decision-making behaviour on the basis of knowledge co-production (Fig. 5). This strategy relaxes or completely deletes old rules for decision-making and facilitates exploration and innovation. In the absence of known solutions, the diversity of actors engaged in the search process and the variety of approaches and strategies is increased. Because of the unknown targets, this strategy is process driven and can be guided by an improved understanding of how complex systems behave[26] and how the human brain works for decision making under uncertainty[27].

Because of human ambiguity aversion (Ellsberg 1961) calculating risks is generally more popular than dealing with uncertainty. Understanding responses to uncertainty, however, better reflects the complex realties of urban life and promises critical advances in science and society. Whereas risks can be assessed and calculated with respect to particular targets, uncertainty is associated with complexity and moving targets.[28] Whereas risks can be quantified and calculated and responsive behaviour can be optimised, addressing uncertainty challenges society by the need to agree on behavioural, moral and ethical rules. This programme will improve our understanding of response mechanisms to uncertainty in complex urban systems, in particular by taking a systems approach to knowledge production and by taking account of diverse knowledge domains.[29]

An important criterion for the policy relevance of research findings on urban health and wellbeing—from a systems perspective, is the systemic role of a variable

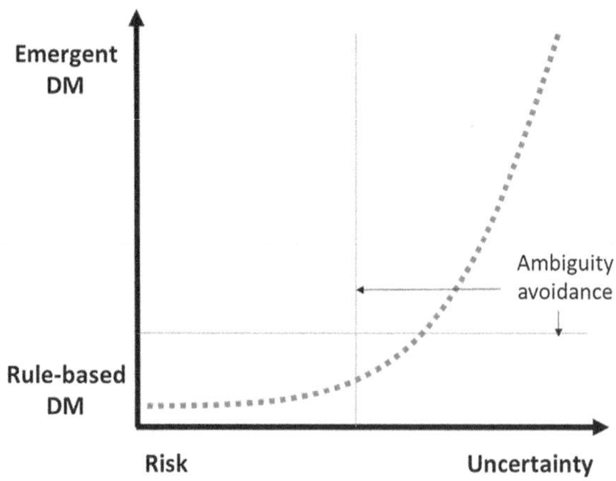

Fig. 5 Decision making (DM) strategies under uncertainty. Because of ambiguity avoidance, decision-making strategies tend to be pushed towards rule-based strategy. *Note* The type of decision making in the *bottom left corner* of this diagram tends to overlap with that of the *upper left corner* in the previous diagram

or set of variables within the urban system. Policies for improving urban health and wellbeing are likely to address issues which have a corrective impact on a set of interrelated variables, rather than only on single variables and making systemically less relevant changes.

Systemic relevance is determined by the role of variables determinint the system play. Their systemic role is determined by the type of interconnectedness or connectivity. Connectivity in turn is determined by:

(1) the number of connections
(2) the strength of connections
(3) the direction of connections
(4) the internal vs. external connectedness

Accordingly, four broad categories of connectedness, which all overlap, can be distinguished, which react differently to policy changes (Vester 2007):

1. **Active** variables: They effect other variables (either more often or stronger) than they are effected by others. Policies which intend to change active system variables can be cost effective and efficient, as one change will automatically cause further changes to connected variables according to the network structure, the number and strength of relations.
2. **Reactive** variables: They are effected by other variables more often than they affect others. Reactive variables are good sensors for the condition of the entire system because their value is largely determined by other system variables. The sensitivity of these variables towards changes in the system is relatively high. Changing or introducing reactive variables into the system can be a useful for source of information on the state of the entire system.
3. **Critical** variables: They are highly interconnected within the system, either by effecting many others or being effected by many others. Changing critical variables can only slightly have an (positive or negative) impact on the entire system.
4. **Buffering** variables: They are part of the system, however their overall interconnectedness in terms of effecting others and being effected by others is relatively low. Changing them can have a positive systemic effect in terms of absorbing tensions in light of overall uncertainty. They are useful levers for creating systemic resilience and avoiding 'business as usual' or 'shying away from decision-making' strategies.

A common belief is that systems should be resilient in order to better buffer risks and recover from shock. The belief is that "… once the system is appropriately compartmentalised … disturbance or risk is more easily countered." Despite its current popularity, resilience can be, but must not always be, the central aim of building systems that work for urban health and wellbeing. Resilience, understood as governance style can also be detrimental to health and wellbeing, especially when it is achieved by specialisation, diversification, segmentation or compartmentalisation. The banking and financial system, for example, is keen on achieving resilience by recovering from disturbances of the financial market and continue with business as usual.

However, it also needs to be asked at which cost the resilience of each system is achieved at the expense of the functions of other systems, which may be as or more relevant for urban health and wellbeing. Basically, what are the opportunity costs of resilience? Sometimes it is better to allow systems to restructure and reorganise instead of rebuilding them and making them inert to change. Such decisions cannot always be taken on what is (scientifically) known to be better. Those decisions are part of experimentation, experience and learning and societal deliberation processes. They cannot simply be calculated based on indisputable facts.

5 A Systems Approach to Urban Health and Wellbeing

5.1 Why a Systems Approach Is Needed

The ways in which cities are perceived and conceptualised, define the types of problems which need to be addressed and the scientific methods applied to address them. Cities can be perceived as systems which can and need to be planned. They also need to be operationally optimized where the built environment and physical spaces change in response to economic and social demands. This view of cities sees them as complicated structures which can be optimised to function according to a set of performance criteria. Problem-solving strategies applied to them have its origins in control theory (Bellman 1964).

Alternatively, cities can be perceived as overlapping social and infrastructural networks in space and time which show open-ended evolutionary processes in terms of social links, connectivity and spontaneous order (Bertaud 2014a, b). This view sees connectivity per person and social inclusion as the key to socioeconomic performance in cities (Bettencourt 2013).

A system can be defined as a complex and adaptive structure, composed of living and non-living parts, interacting in networks, establishing networks of interacting components and behaving as a coherently organized entity. Cities can be perceived as complex adaptive systems, with a number of characteristics described by Sanders (2008) and composed of subsystems which are also complex, adaptive and embedded into the overall supporting global ecosystem.

Hancock and Duhl (1988) define cities as "the example par excellence of complex systems: emergent, far from equilibrium, requiring enormous energies to maintain themselves, displaying patterns of inequality and saturated flow systems that use capacity in what appear to be barely sustainable but paradoxically resilient networks".

"Complicated" or design-and-control problems require different approaches than "complex" or wicked socio-economic problems of urban health and wellbeing (Fig. 6). Both appear in the design and management of complex systems. Complicated problems are characterized by linear causality and feedback-control loops that hold a system within an optimal range. Objectives are clear, measurable and the mechanisms are in place which lead to corrective actions or adjustments in

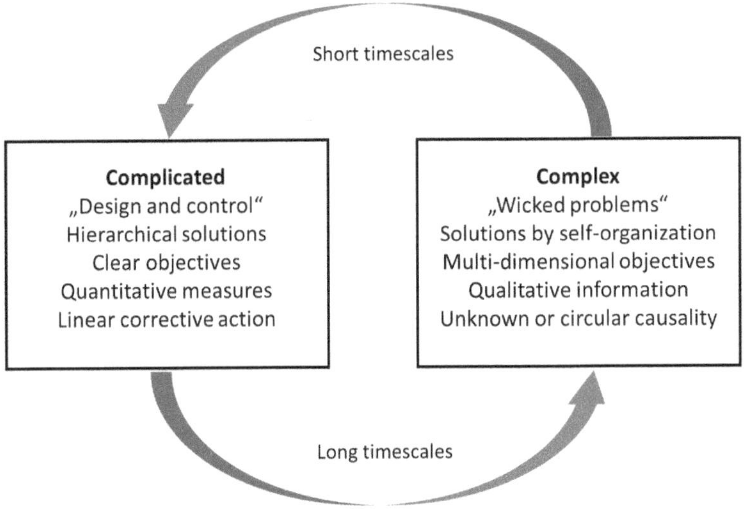

Fig. 6 Types of problems in complex systems of urban health and wellbeing (Modified from Bettencourt and Brelsford 2015: 195)

time. "Such problems can be defined through optimal design and often lead to hierarchical solutions ..." (Bettencourt and Brelsford 2015: 195).

Complex problems are problems with circular causality, gradual evolution, multidimensional and changing objectives and qualitative information. They typically refer to questions of sustainability, ecology and human development and they are characterized by radical uncertainty. Strategies for solving those problems are not readily in place and often emerge by processes of self-organisation. The information needed to solve those, often societal problems, is distributed widely and needs to be pooled and coordinated in order to be useful. Instead of optimal solutions, path dependence and incremental improvements are the norm (ibid: 196).

Wicked problems are, for example, palliative care problems which are widely prevalent in our health care system. They make it difficult to decide about ending lives or prolonging mechanically supported, painful processes of dying. Issues associated with older people's care and physical and mental co-morbidities are typically complex to manage (Periyakoil 2007).

Different types of knowledge and strategies are required to address those different types of problems in cities. Bertaud (2014b), for example recognizes that engineered planning and spontaneous development both have a place in urban development and he asks when top-down urban planning should give way to allow for spontaneous order to emerge for improved housing availability in cities. His research has revealed that sometimes too much planning can be damaging when, because of regulations, consumption is imposed on consumers who cannot afford living and building according to the imposed regulations (e.g. minimum house size). The negative consequences of those regulations are social exclusion and slums or wasteful use of urban space for housing[30].

There are also positive examples of spontaneous housing, for example, when informal settlements are not illegalised, allowed to evolve to the affordability of individual households and connected to public service infrastructure for water, sanitation, security, energy and waste management. The quality of life of people living in informal settlements[31] is defined by their capabilities, which are not restricted by rules such as minimum consumption standards in housing markets and enhanced by providing access to public infrastructure such as electricity grids, water supply and garbage collection.

Similarly, despite the widespread idea among urban planners that urban sprawl needs to be contained, technological innovation has encouraged urban dispersal and allowed lower income classes to connect to the city. Whether urban sprawl creates private and social costs or benefits largely depends on the diverse characteristics of cities as complex systems. In some cases, sprawl needs to be regarded as features of economic diversification. Compact cities with all employment located in a central business area, limit economic opportunities while sprawled cities offer firms and households more choices and in combination with technological innovations reduce congestion and do not necessarily worsen pollution (Kahn 2006).

Where to strike a balance between planning and spontaneous order, or top-down versus bottom-up (Easterly 2008), is a continuing challenge for governance and is most likely a question of the costs and benefits of regulation versus spontaneous order and the values attributed to those costs and benefits at a particular time and location. The plural values which define people's health and wellbeing in multiple dimensions of urban life[32], are invisible drivers of the process of urban change. The constant imbalance and distance from an optimum or preferred state for multiple trade-offs (for example between commuting distance and land consumed per household, or air quality and the use of private cars), prevents the urban system from stagnation in equilibrium and drives constant, innovation, destruction and renewal.

Hayek (1945) noted the impossibility of central planners to grasp enough of the widely dispersed and detailed local knowledge which is critical for making decisions for local adaptive change. If science-based decision making is asked for in society, there need to be ways to integrate diverse knowledge domains and overcome communication barriers between science and society[33]. This dilemma needs to be recognized in order to achieve good governance[34]. Depending largely on technological and social progress enabling the making of informed decisions for the benefit of human health and wellbeing, there is likely to be no universal solution to this dilemma of organization and freedom (Ostrom 1983)[35].

Recognising both types of decision processes (top-down and bottom-up) and types of urban organisation and development (planned versus self-organized), two broad types of knowledge are embraced by this programme. On the one hand, knowledge which has been termed "scientific" and is presumed to be precise, based on clearly defined features of rationality and assuming that a sufficient amount of quantitative data allows optimal decision making. On the other, knowledge based on experience, tradition and qualitative information, also referred to as "extended facts". "These (extended facts) may include craft wisdom and community knowledge of places and their histories, as well as anecdotal evidence, neighbourhood surveys, investigative journalism and leaked documents" (Funtowicz and Ravetz 2003).

The rationale for embracing both types of knowledge is that the quality of solutions suggested by science is insufficient because complexity and value-loading are excluded from "pure" scientific approaches to problem solving results[36] (Fig. 7). Rationality is not universal. Different types of rationality belong to different types of knowledge, which again are more or less relevant in different socio-ecological contexts which can be simple or complex. The issue is one of choosing the appropriate method of investigation for the nature of the problem. Deliberative methods applied in the coproduction of knowledge are based on different assumptions with regards to the nature of the good and the interaction capacities of individuals. Methods are the rules for how to produce science. They define who should participate in providing data, what is considered data, which form data should take and how to reach a conclusion (Vatn 2009).

The coproduction of knowledge is an important mechanism to ensure the quality of science-based decision making. Why do we need to co-produce knowledge? Knowledge of complex urban systems needs to be co-produced because the urban development problem is a social coordination problem. Bettencourt et al. (2014) explain, that "obtaining and coordinating information across levels of organisation is an archetypal open-ended coordination problem: whenever individuals have common interests, or goals, and their actions depend on actions of others, they must coordinate their actions in order to reach their goals (Schelling 1960). But recognising that their goals and actions interact, is often not obvious, and arriving at this recognition, and devising ways to coordinate their problem-solving actions, requires information and deliberation."

The purpose of co-producing knowledge is also to make use of different types of knowledge domains which can address different types of problems we are facing in

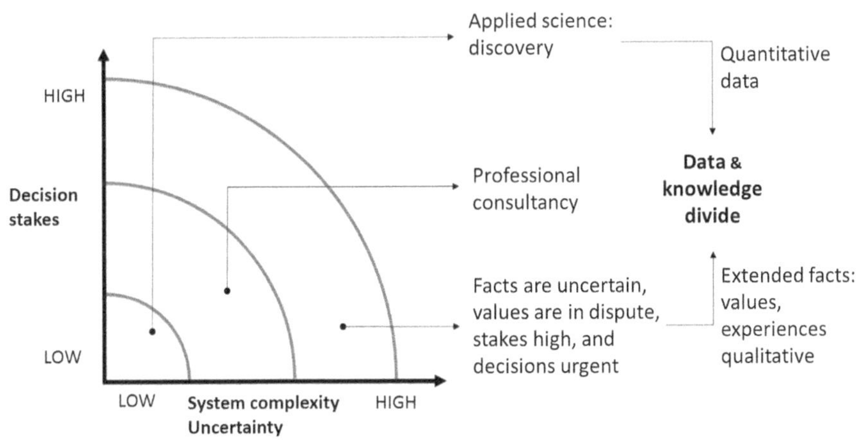

Fig. 7 Bridging the data and knowledge divide by different types of knowledge. Image based on a diagram by Funtowicz and Ravetz (1993)

complex urban settings. Organising the co-production of knowledge in ways so that problems which emerge in complex systems can be better solved than without making use of multiple types of knowledge, can also be regarded as a step towards creating organizational[37], systems[38] or collective intelligence[39]: different types of knowledge need to be communicated and coordinated in order to produce quality decisions at higher organisational scale which better respond to the degree of uncertainty, complexity and decision stakes involved. Citizens need to be invited into the decision-making process, for example by virtual models of a city, allowing collective visualisation and planning. Collective intelligence means not only higher degrees of connectedness but also more sophisticated systems of information processing and foresight which will be created by deliberative and participatory research and knowledge co-production[40].

5.2 *What Is a Systems Approach to Urban Health and Wellbeing?*

The systems approach to knowledge creation taken by this programme applies to science and to society. Our systems approach defines the city as a complex adaptive system and recognises the multiple values generated by the urban system for the health and wellbeing of its inhabitants. Values, also intrinsic values, of urban system goods and services are socially constructed by a demand for them. Vice versa, there needs to be a demand for avoiding disservices such as air, noise, and light pollution, or exposure to risk in urban traffic. Multiple and plural values of urban system goods and services then need to be institutionalised, e.g. by laws, regulations and incentives mechanisms, for actual behavioural change to occur for improving urban health and wellbeing (Brondizio and Gatzweiler 2010).

Systems approaches in science go beyond the recognition that health is effected by distal factors or factors defined at multiple levels of organisation. They explicitly allow for dynamic processes including feedbacks as well as interdependencies and interactions between individuals and environments over time. These feedbacks and interdependencies can result in non-linear relations and unanticipated effects distant in space or time. In going beyond the traditional emphasis on isolating the “independent” effect of specific factors to an understanding of the functioning of the system as a whole, the use of systems approaches implies a paradigm shift in the way in which population health is conceptualised, studied, and intervened on. Puppim de Oliveira et al. distinguished four elements of systems approaches, relating to analytic method, interdisciplinary, transdisciplinary, and scaling/bounding (Puppim de Oliveira 2015).[41]

For society, a systems approach means co-producing knowledge for urban health and wellbeing in collaboration with science. This means recognising how different urban system functions and modes of urban life are connected to particular health and wellbeing outcomes, raising awareness and education on interrelated issues of

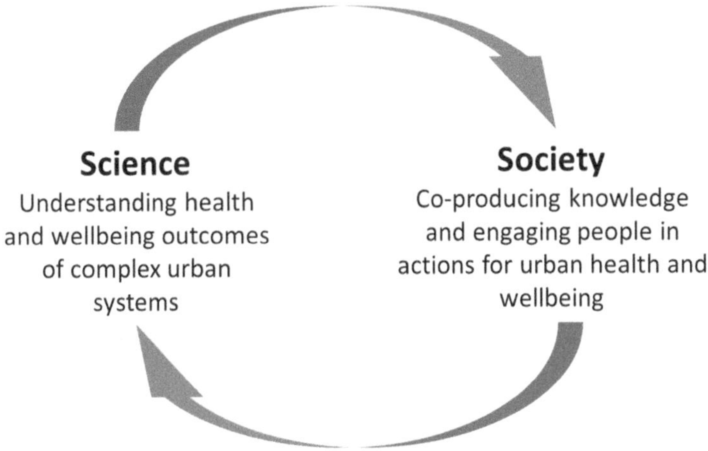

Fig. 8 The programme's systems approach to knowledge creation

health and wellbeing, creating demand and opportunities for entrepreneurship, business and civil society engagement for health and wellbeing, creating networks of like-minded system thinkers and being agents of change for improving health and the quality of life in cities. People are the agents of urban change[42]. A systems approach to society recognises people as agents who have the right and freedoms to change cities for their health and wellbeing (Mitchell 2003).

The programme recognizes the limits of purely science-based decision making and therefore aims at engaging societal stakeholders for co-creating knowledge of urban health and wellbeing (Fig. 8), by making them part of the research process (e.g., participatory or companion modelling) and supporting communication and deliberation activities between science and society (e.g. by organising science policy dialogues and facilitating networking, education and capacity building).

Given the novelty of systems approaches in the study of population health and the city, the programme takes a broad view of systems approaches (Box 1). Systems approaches involve one or more of the following elements[43]:

(1) the development of new conceptual models of the processes leading to health in urban settings that incorporate dynamic relations. These conceptual models must be specific to a given research problem or question; the development of these models may involve input from stakeholders as well as scientists as appropriate to the research problem and context;
(2) the use of systems tools and formal simulation models, such as agent-based models, systems dynamic models, or other systems modelling tools to better understand the functioning of the integrated urban and health system or predict changes to health under various hypothetical interventions;

(3) the integration of various sources and types of data (including spatial, visual, quantitative and qualitative data) in the conceptual models and/or the formal simulation models as well as the identification of important data gaps that need to be filled in order to advance understanding of how the system works.

Box 1: A systems approach to urban health and wellbeing

A systems approach applied to urban health and wellbeing taken by our programme

- perceives the city as a complex adaptive system which functions to provide a range of goods and services enabling people to live urban lives they value.
- integrates human health concerns into urban system function (structure and processes).
- addresses knowledge creation in science and society in order to harness urban complexity by solving complicated and complex types of problems and the different types of knowledge needed therefore.

For society, a systems approach means co-producing knowledge for urban health and wellbeing in collaboration with science, recognizing how different urban system functions and modes of urban life are connected to particular health and wellbeing outcomes, raising awareness and educating on interrelated issues of health and wellbeing, creating demand and opportunities for entrepreneurship, business and civil society engagement for health and wellbeing, creating networks of like-minded system thinkers and being agents of change for improving the health and quality of life in cities.

For science a systems approach means

(1) the development of new conceptual models of the processes leading to health in urban settings that incorporate dynamic relations. These conceptual models must be specific to a given research problem or question; the development of these models may involve input from stakeholders as well as scientists as appropriate to the research problem and context;
(2) the use of systems tools and formal simulation models, such as agent-based models, systems dynamic models, or other systems modelling tools to better understand the functioning of the integrated urban and health system or predict changes to health under various hypothetical interventions;
(3) the integration of various sources and types of data (including spatial, visual, quantitative and qualitative data) in the conceptual models and/or the formal simulation models as well as the identification of important data gaps that need to be filled in order to advance understanding of how the system works.

It is expected that the use of systems approaches will advance our understanding of the determinants of urban health and wellbeing as well as enhance our ability to intervene to improve health and reduce health inequalities in urban settings. The use of these approaches will yield new insights into old problems, identify new important questions, and point to fresh kinds of data that need to be collected or processed in order to advance the planning and management of healthy cities.

In the process of coproducing knowledge for urban health and wellbeing, the programme will embrace multiple types of knowledge. The **T-shaped knowledge management approach** recognizes the value of in-depth, disciplinary knowledge and broader context sensitive interdisciplinary knowledge (Brown et al. 2015). The programme's activities are defined to enable scientific exchange between and among disciplinary fields of science to achieve knowledge gains from joined-up thinking and to communicate and develop those knowledge gains in collaboration with stakeholders[44] from society in order to accomplish transdisciplinary knowledge creation (Fig. 9). The programme aims at building on the achievements made in interdisciplinary knowledge generation. According to Lawrence (2015) they include:

1. development of conceptual and analytical frameworks
2. a diversification of methods and approaches in precise localities,
3. specific cases showing the creative, reflexive and transformative capacity of transdisciplinary inquiry, and concerns about the asymmetries of power and control of participants during processes of the co-production of knowledge.

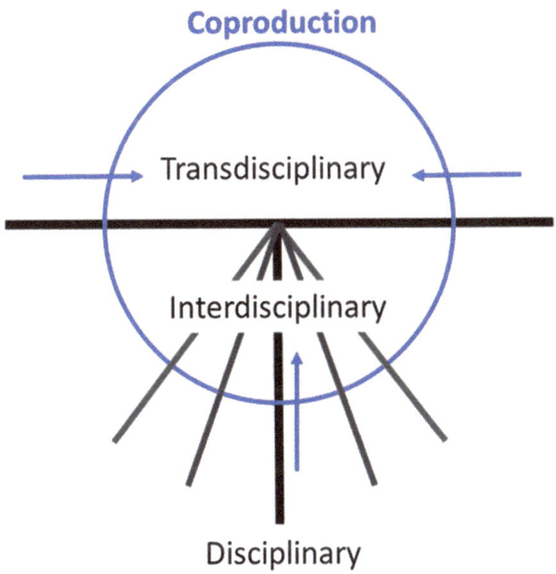

Fig. 9 T-shaped knowledge management for the coproduction of knowledge for urban health and wellbeing

Inter- and transdisciplinary research on urban health and wellbeing may mean including various scientific disciplines such as architecture, urban planning, engineering, law, but also representatives of other urban sectors as "education, housing, criminal justice, and transportation; social movements and activists from environmental justice, women's health, food justice, occupational safety and health, and human rights; policymakers from local, state, national, and global governance bodies; and residents of the communities most affected by health problems. Each of these stakeholders has a role to play in conceptualising, implementing, interpreting, and disseminating research studies." (Freudenberg et al. 2009: 13).

6 Implementing the Systems Approach: A Conceptual Framework

In the following we develop an integrated conceptual model which helps structure the ways in which we think about cities as complex systems and how they function for people's health and wellbeing. The development of integrated conceptual models applied to urban ecosystems is being advanced by several scholars (Costanza 1996). We do so by merging the ecosystem approach to cities with the capabilities approach developed to understand quality of life.

Urban Health

Human health is defined as a state of complete physical, mental and social well-being and not merely by the absence of disease or infirmity[45]. Urban health, in turn, refers to healthy human lives in cities which function to maintain human health. Hancock (undated, 2015) has suggested the following definition of a healthy city: A healthy city is "one that is continually creating and improving those physical and social environments and expanding those community resources which enable people to mutually support each other in performing all the functions of life and in developing to their maximum potential."

Human health and wellbeing in cities as an aspired goal of urban life, recognizes the many different direct and indirect determinants influencing the quality of life in cities. Human and urban health mutually co-determine each other and are in constant flux. Human health and wellbeing also co-determine each other.[46] In contrast to states of human health, wellbeing is not merely about income, it describes a situation in which people are free to choose to do and be what they value (Sen 1999).

Urban Systems

In the following we will refer to cities as "urban systems"—not to be confused with urban ecosystems (McPhearson et al. 2014; Gómez-Baggethun et al. 2013) which are frequently referred to as the natural capital of cities, comprising of "urban green": parks, gardens and other urban vegetation, including animals and insects that inhabit urban green spaces. Urban systems include humans and comprise of man-made, social, financial and natural capital. Research continues to establish the evidence for the fact that humans and nature are part of the same biological continuum.[47] The services provided by urban systems are therefore provided by and for the city's human residents.

Grimm et al. (2011) define the following attributes which make cities "end-member ecosystems":

(1) they are heterotrophic and extremely energy intensive
(2) they therefore require large inputs of energy and materials
(3) they produce copious amounts of waste compared to most ecosystems and often lack effective assimilation mechanisms to handle these wastes (or strain existing ones)
(4) urban ecosystem function is controlled not just by biophysical factors but also by social and political forces (although this type of control now affects most ecosystems to some extent, it affects cities in a profound manner)
(5) one keystone species-humans—exerts overwhelming control on (urban) ecosystem processes.

Urban systems can be defined as an "open, fully contained, and dependent, growing subsystem of the materially closed, non-growing ecosphere." (Rees 1995; Daly 1992). They are social-ecological-technological dynamic complexes of living and nonliving components, interacting as *functional units*.[48] The social, natural and man-made environments of urban systems overlap in creating the urban environment and they are contained in natural ecosystems outside their boundaries, providing them with water, air, materials and energy. The difference between urban systems and natural ecosystems is that most of the goods and services provided in urban systems stem from secondary production and are made and consumed by humans[49], who are heterotroph, while ecosystems consist of primary (autotroph) and secondary producers. Urban systems exchange and transforms materials, substances and information, and therefore the concept of metabolism[50] can be applied to cities. Economic production in cities needs to be viewed as ecological consumption by heterotroph entities[51], feeding on their environment. Cities and human economies differ from primary producers (e.g., plants) who feed on low-grade energy sources (e.g., sunlight) and therefore, "... cities per se constitute only a small part of the ecosystem complex needed to support urban human populations" and therefore do not strictly qualify as ecosystems (Rees 2003: 124).

Urban System Structure and Processes

Urban systems constitute elements of natural ecosystem structure, such as species composition and biodiversity, built structure (houses, roads), designed structure (landscape architecture and urban design), as well as social and cultural structure (demographic composition, social networks, institutions, economic and political systems). Urban system structures facilitate the different types of processes for maintaining urban metabolism[52]—in terms of energy equivalents and in terms of a city's mass fluxes (flows of nutrients, water, materials). The concept of social metabolism is similar, however it would emphasize flows of information, institutions and plural values attached to material and energy flows. Structures and processes of urban systems here are also referred to as urban system function. The city performs a wide range of functions which are defined by how its structure and processes are designed, organised and governed.

Urban System Services

The categorisation of urban system services we suggest here is similar but not identical with previous categorisations of ecosystem services, such as in the Millennium Ecosystem Assessment (MA 2005), or The Economics of Ecosystem Services and Biodiversity (TEEB 2010, 2011). Because we define cities as complex, dynamic social, ecological and technological systems, the services they provide are not merely restricted to those provided by natural capital; they also include services made by and for humans.

Supporting and habitat services are fundamental for all other ecosystem services. They make physical space and infrastructure available for life in the city. Green (e.g. parks), blue (e.g. lakes), and grey (roads and buildings) infrastructure and space provide basic life support services, like shelter, waste management, water treatment and sanitation, production of goods and energy provision. The habitat functions make the city a liveable place by providing the space, hardware and infrastructure which enable material, energy and data flows—thereby facilitating urban metabolism.

Provisioning services include goods and services provided by urban system, some of which can be exchanged on markets, provided by the public or co-produced. Goods include e.g. processed and unprocessed food, water, manufactured goods, medicines, computers, or books. Most of the unprocessed food products come per-urban areas. Services include access and use of roads, communication and other public infrastructure, security, waste management, health care and education systems, disaster response and emergency systems. The provision and production of goods and services can be organised publically, privately or in combinations of both. Note that in our categorisation, the *availability* of such infrastructure would be considered as a supporting service.

Regulating services include the benefits obtained from regulation mechanisms in urban systems. They include regulation mechanisms of the urban green spaces, including regulation of climate and temperature, as well as human immune system regulation by exposure to microbiota. Regulation services also include health benefits derived from having a system of rules and regulations (institutions) in place by means of which the urban system is governed (in the social urban space) and managed (in the economic and technological urban space). Institutional infrastructure regulates social interactions and other urban metabolisms, e.g., regulating access to public places and services, markets and businesses, traffic, energy and water supply, the collection and use of data, or food safety by environmental and quality standards.

Socio-cultural services are the non-material benefits people obtain from urban system functions—they emerge from all other services of the urban system and include aesthetic experience, recreation, creativity and innovation, spiritual and inspirational enrichment, cognitive development, and social relations which emerge in social space and liberties for economic, social, cultural and political innovation. They include the emergence and exchange of ideas, creativity from exposure to cultural diversity and different forms of cultural expression, recreation and leisure.

Emergence in a social system is a product of interactions among different agents. The behaviour of one observed agent is the result of others s/he is interacting with, which again is the result of yet further interactions with other sets of actors. These interactions are nonlinear, i.e. they are more than the sum of a systems parts. The summing up of the constituent parts would not explain the behaviour of the whole system (Holland 1998).

In the case of a city, the interactions of the subsystems (including the human population) would lead to patterns of wellbeing and illness which are difficult to predict precisely (i.e. unintended consequences). The system adapts to the environmental changes through processes of self-organization and evolution.

Despite highly suggestive by some authors (Geddes 1949; Batty 2003) argues that there is no (top down) invisible force or hidden hand guiding the physical, spatial and social organization of cities. Rather, the patterns that can be observed in the city emerge from a myriad of interaction types and decisions from the bottom-up. Cities and urban space will therefore grow as a form of adaptation. This is what happens in the enlarging peri-urban spaces. Even within the geometrically defined central parts of the city, there are modifications, adaptations that are introduced by those living in these spaces.

The relevance is that we have to promote local responses and flexibility to respond to changing environments at local levels. We cannot depend on responses designed at the centre. Coproduction of knowledge is an example of developing local responses.

Our conceptual framework (Fig. 10) aims at explaining how urban health and wellbeing is created and can be changed by the structures and processes of the urban system and the services it provides. It is one way to conceptualise the city as a complex social, ecological and technological system. Recognizing multiple types of knowledge, as we do in this programme, does not exclude different conceptualisations of the city as a complex system in which people are provided the opportunities to achieve certain levels of health and wellbeing.

Urban system services are co-evolving, emergent and mutually enforce each other in positive and negative feedback cycles. An intact road infrastructure, for example, invites people to drive their cars, which over time and with increasing traffic, demolishes the roads and if unregulated and not maintained, will cause congestion, pollution and stress. The use of cars can also contribute to a reduction in physical activity and downstream lifestyle diseases such as obesity and hypertension. Urban system functions are therefore not provided at all times, to the same extent and quality and not equally to all urban inhabitants.

The connectedness of urban systems, i.e. the quality of various urban infrastructures which enable interaction, communication and exchange, is an essential determinant for the performance and functioning of urban systems. Connectedness

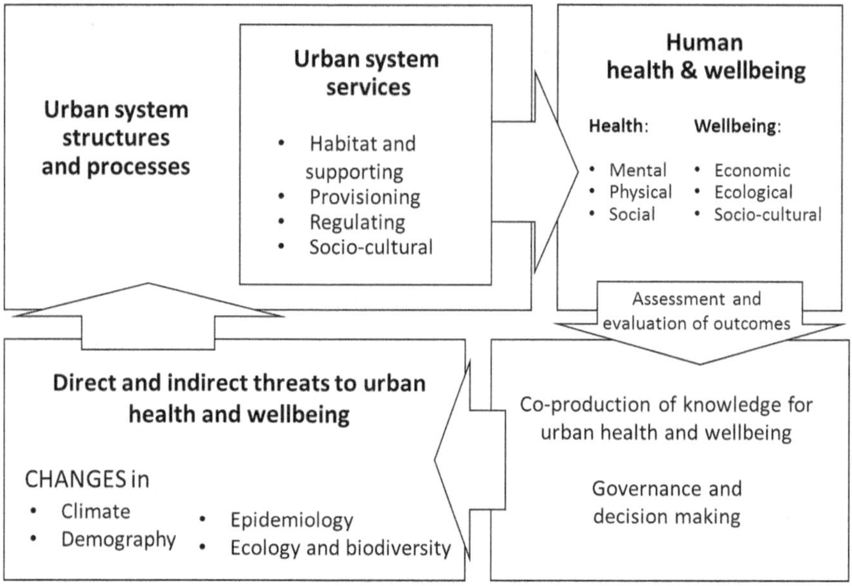

Fig. 10 A conceptual framework for a systems approach to urban health and wellbeing

refers to the strength of internal and external relations of the system variables. In a state of low connectedness system variables are dominated by outward relations and effected mainly by external variability. Systems with a high degree of connectedness are dominated by variables which affect each other inside the system more strongly than they are effected by variables outside the system (Gunderson and Holling 2002). The relationships of system variables to each other and how they influence each other effects the behaviour of the system. These different types of system behaviour based on the interrelatedness of system variables, results in higher or lower degrees of connectedness (Tables 1 and 2).

Urban Wellbeing

Next we need to ask how urban system functions (structures and processes) affect the performance of urban systems with respect to health and wellbeing. Wellbeing describes a situation in which people are free (or enabled) to choose to do and be what they value. In Sen's (1999) theory of "development as freedom" the concepts of "functions and capabilities" match with our ecosystem approach and urban system functions. The difference is that we now shift from the urban system, which provides a range of goods and services, to the human agent who chooses a range of human enabled opportunities within the urban system.

The urban system services represent a kind of choice set from which the human agent chooses to be and do what he values or has reason to value being and doing. Valued functionings of the human agent comprise of material (e.g. to be well

Table 1 Urban system services and examples of health benefits and risks

Type of urban system service	Description of urban system services	Examples of health benefits and risks
Habitat and supporting	Physical space and infrastructure for living and working in the city. Green (e.g. parks), blue (e.g. lakes), and grey (roads and buildings) infrastructure and space are created to provide basic life support functions, like shelter, waste management, water treatment and sanitation, production of goods and energy provision. The habitat services make the city a livable place by making the hardware and structures available which enable material, energy and data flows—thereby facilitating urban metabolisms	Housing-related health risks • Allergies and asthma as a result of accumulation of indoor pollutants and dampness • Spread of infectious diseases • Microbiome of the built environment increases or disrupts immune system regulation, exaggerate or suppresses inflammation • Respiratory and cardiovascular diseases from indoor air pollution • Risk of airborne infectious diseases because of inadequate ventilation • Illness from temperature extremes • Risks of home injuries • Health benefits from urban green space • Reduction of urban heat • Offsetting of greenhouse gas emissions • Attenuating storm water • Providing urban residents spaces for physical activity and social interaction • Exposure to microbiota which educates the human immune system
Provisioning	Goods and services provided by the urban system some of which can be exchanged on markets, provided by the public or co-produced. Goods include e.g. food, water, manufactured goods, medicines, computers, or books. Services include e.g. access and use of roads, communication and other public infrastructure, security, waste management, health care and education systems, disaster response and emergency systems. The provision and production of goods and services can be organised publically, privately or in combinations of both	• Unhealthy urban food environments are associated with increased obesity • Urban farmers markets increase healthier food choices • Access to pharmaceuticals and medicines • Access to public health facilities • Transport and communication infrastructure can improve social networks • Social determinants of health • Health insurance • Hospitals • Clean water • Sanitation facilities

(continued)

Table 1 (continued)

Type of urban system service	Description of urban system services	Examples of health benefits and risks
Regulating	Benefits derived from having natural regulation mechanisms and man-made system of rules and regulations in place by means of which the urban system is governed (in the social space) and managed (in the economic and technological space). Institutional infrastructure regulating social interactions and other urban metabolisms, e.g., regulating access to public places and services, markets and businesses, traffic, the collection and use of data, food safety and environmental standards in the urban economy	• Climate and temperature regulation by urban vegetation and technology • Air purification, run-off mitigation, decomposition of waste, noise reduction • Human immune system regulation by exposure to microbiota • Policing and public safety • Enforcement of traffic rules and road safety • Food safety standards and controls • Disease control regulations • Hygiene regulations/standards • Medical law • Public health law • Construction regulations • Standards in the Control of Hazardous Substances • Environmental regulations
Socio-cultural	Benefits created in urban socio-cultural spaces. Social space and liberties for economic and political innovation, exchange of ideas, creativity from exposure to cultural diversity and different forms of cultural expression, recreation and leisure, space for spiritual enrichment, art and cognitive development. E.g. cultural events, religious places, "Heimat" (sense of belonging), exhibitions, libraries, cultural heritage values (e.g., historical places), cultural diversity	• "If the role of cultural systems of value in health is ignored, biological wellness can be focused on as the sole measure of wellbeing, and the potential for culture to become a key component in health maintenance and promotion can be eroded." (Napier et al. 2014: 1607) • Cultural diversity in healthcare system increases inclusion of minorities • Cultural competence can improve quality of healthcare; culturally adapted healthcare can improve patient understanding and health outcomes • Culture's dictation of female and male roles that limit women's mobility and ability to seek health care

nourished or to be well equipped) and non-material conditions and actions (e.g. to be wise and educated) and they relate to many different dimensions of urban life: health, employment, social capital, self-realisation and culture—each being provided for by urban (sub)system functions.

Table 2 Examples of urban subsystems and the goods and services they provide

Subsystems	Goods and services
Energy Water Food Roads Housing	Providing energy, water, food and physical space for movement and living/shelter is a basic requirement for people to live in cities. The quality of the provision of these goods and services determines strongly determines health and wellbeing of people living in cities
Waste management	Waste management infrastructure and services
Traffic	Providing transport infrastructure and regulation
Social security	Social protection programmes and safety nets supports the poorest
Public health and sanitation	Hospitals, medical personnel and health & sanitation infrastructure
Legal	Jurisdiction, courts and accountability
Law enforcement	Police, policing and security
Administration, finance and accounting	Banks and other financial organizations facilitate the flow and regulation of financial resources
Education and research	Schools and higher education facilities provide knowledge
Emergency and disaster	Urban emergency plans, urban disaster risk management plans, risk assessments and projections
Telecommunication	Telecommunication and internet infrastructure enable connectivity, communication and exchange
Employment	Employment agencies aim at matching supply and demand
Economy, trade, industry and business	Providing space and opportunities for economic activity, securing benefits and protecting consumers
Political system and governance	Providing the infrastructure and procedures for political decision making in cities ("city making")
Urban green	Urban green areas are places for mental and physical recreation. They also provide basic supporting functions like air, shade, temperature regulation and water table balancing
Culture, tourism and recreation	Cities offer touristic attractions and places with historical, heritage and recreation value

Capabilities are the freedoms which enable agents to realise their valued functionings. The concept of capabilities strongly relates to that of the "enabling institutional environment."—which is a regulating and provisioning function provided by the urban system. Having capabilities enables people to actually make a choice about what they want to be or do and having real opportunities with regards to the functionings they aspire, value and have reason to value.

There is a constraining and enabling component in every institutional environment. Therefore, freedom, here should not be misunderstood as "free from any restraint" but rather as enabling drivers for achieving what people value being and doing. Having fewer choices can sometimes be more enabling (and thereby contributing to wellbeing) than having to make (costly) trade-offs among many choices. The central idea is to increase freedoms and not increase the number of choices an agent has. The institutional environment automatically creates boundaries as to which and how choices can be made and who is included. Willingly refraining from a functioning a person could realise but does not value, also counts as increasing freedom.

Figure 11 illustrates how urban system services enable aspired human functionings. Increasing overlap of enabling urban system services and aspired functionings results in increasing capabilities and higher levels of health and wellbeing which in turn either leads to intentional design of interventions or emergent evolution of self-organised order which facilitate change in policy and action. Change for increasing capabilities comes with social and technological innovation. As urban systems change towards a higher overlap of the enabling system services and human functionings (what people aspire), they improve urban health and wellbeing.

In our conceptual framework, change is not only unidirectional. As explained with reference to the adaptive cycle introduced by Gunderson and Holling (2002), after reaching a critical level of connectedness, systems can collapse, rebuild and reorganise. The trigger for renewal in social or technological networks can come from inside or outside the system and occurs when the benefits of established interaction patterns no longer outweigh the costs of interacting differently. Renewal is essential for creating new space for new interaction patterns and different, maybe better adapted types of system behaviour. As the conceptual framework includes complexity and uncertainty, specific performance indicators cannot be defined. How successful people are in achieving aspired levels of urban health and wellbeing will be determined by the level of understanding of how complex urban

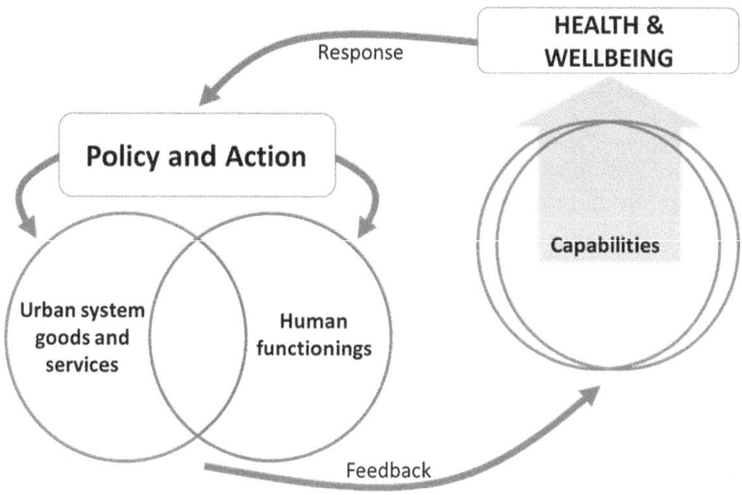

Fig. 11 Increasing overlap of enabling urban system services and human functionings leads to increasing capabilities and higher levels of health and wellbeing

systems work, which role and scope the human actor has in actively modifying it and how successful people are in coordinating and converging knowledge across sectors and disciplines and thereby upscaling collective intelligence.

Recent research findings on the dynamics of complex systems and human development recognises the simultaneous evolution of urban infrastructural networks and social structures and services and thereby strongly supports our conceptual approach of combining urban system functions and human functionings (Holston 2008; Jacobs 1961; Bettencourt et al. 2007; Andris and Bettencourt 2014).

7 Vision and Goals of a Science Programme on a Systems Approach to Urban Health and Wellbeing

Human health had been identified in ICSU's Strategic Plan 2006–2011 as a new research priority, with the stated goal "to ensure that health considerations are duly taken into account in the planning and execution of future activities by building on the relevant strengths of Scientific Unions and Interdisciplinary Bodies." ICSU's strategic plan 2012–2017 (ICSU 2011a) states the goal to "promote transdisciplinary projects that adopt this framework and aim to generate understanding and products that are useful to policy-makers." In the interdisciplinary science plan (ICSU 2011b) of the ICSU planning group on health and wellbeing in the changing urban environment: a systems analysis approach, the overall goal of the research programme was "to produce and apply models of health and wellbeing in the urban environment that simultaneously advance our understanding of the origins (i.e., causation) of the problems and potential solutions in this area and which can therefore inform decision-makers."

The *vision* of the programme, are cities functioning as integrated complex systems which sustainably provide benefits for the health and wellbeing of its residents. This requires people who are enabled to perform all the functions of life and develop to their maximum potential.

The *goals* of this science programme reflect the structure of the programme as proposed by the planning group report (ICSU 2011b). It aims at

1. Promoting and coordinating research projects

Goal 1 will be achieved by generating high-impact scholarly outputs; by enabling better-informed decision-making by the variety of stakeholders involved in urban-health and wellbeing; and by establishing systems approaches to health and well-being in the urban environment as a thriving and relevant area of interdisciplinary research.

2. Developing methodologies and identifying data needs

Goal 2 will be achieved by proposing innovative systems methodologies and approaches that are applicable to the particular challenges of health and wellbeing in the urban environment; by identifying data needs that inform and influence the various ongoing and planned observation and monitoring initiatives; and by identifying new data from third research studies that can be made openly available.

3. Building and strengthening capacity

Goal 3 will be achieved by building the scientific capacity and supporting interdisciplinary platforms necessary for undertaking research into urban health and wellbeing using systems approach; by building the capacity of policy makers and practitioners to understand scientific research on urban health and wellbeing using a system approach; by facilitating the establishment of networks in research, business and civil society communities which take a systems approach to urban health and wellbeing and by expanding the number of students and young scientists with an interest to undertake research, entrepreneurial and civil engagement activities for urban health and wellbeing using systems approaches that engage with substantive policy relevant issues.

4. Communicating new knowledge: promotion and outreach

Goal 4 will be achieved by creating a virtual forum as a point of reference for the scientific community and other stakeholders; by promoting interaction and collaboration between researchers and with other relevant stakeholders, via conferences and workshops and by making results available to multiple stakeholders in a relevant format that allows them to be easily understood.

8 Guiding Principles and Thematic Areas

The programme will apply four *principles* and engage in four *action areas* and thereby contribute to creating knowledge for the transformation of cities towards urban health and wellbeing. Principles respond to the question of "how" the programme activities will be performed. Action areas respond to the question of "what" types of activities the programme will facilitate.

The *principles* refer to the type of projects for urban health and wellbeing and how they are carried out. The programme will make the attempt to include all types of research which contributes to a better understanding of the components of complex urban systems and their relations to one another. It is inclusive of all relevant fields of science and stakeholders who can contribute to the co-production of knowledge for urban health and wellbeing. Urban sectors will be included, such as transport, energy, health, industry, business, civil society, and government. The principles for how the programme activities will be carried out are therefore: systemic, inclusive, inter- and transdisciplinary, integrative and intersectorial.

In addition, the programme is committed to the principles of open data and open science as developed by Science International in an accord on the values of open data in the emerging scientific culture of big data. The accord "identifies the opportunities and challenges of the data revolution as today's predominant issue for global science policy. It proposes fundamental principles that should be adopted in responding to them. It adds the distinctive voice of the scientific community to those of governments and inter-governmental bodies that have made the case for open data as a fundamental pre-requisite in maintaining the rigor of scientific inquiry and maximising public benefit from the data revolution in both developed and developing countries."[53]

The *thematic areas* of the programme define the fields in which activities will be carried out, as elaborated under the programme goals. The programme will facilitate exchange of knowledge in science and other fields of knowledge, communicate science to policy and the rest of society, develop communicable concepts for different stakeholder groups, build capacities for doing and communicating a systems science for urban health and wellbeing, and the programme will deliver critical insights for improving the governance policy making for health and wellbeing in cities. On the basis of those insights, science donors will be encouraged to promote systems approaches for research on urban health and wellbeing. The thematic areas of the programme therefore encompass knowledge creation, concept development, capacity building and policy and decision support.

The programme activities will support implementation activities by individuals or groups of urban thinkers and activists, institutions and organizations, initiatives and movements, scientists and independently funded scientific research and action projects. The programme will undertake efforts to support implementation activities by

1. promoting a systems approach to health and wellbeing in a changing urban environment among scientists and decision-makers, by engaging ICSU Unions and other scientific bodies, creating networks of scientists and engaging citizens
2. strengthening capacity to conduct research on health and wellbeing in an urban environment using a systems approach, by identifying data gaps and possibilities for collecting data generation
3. advocating for funding of systems approaches for health and wellbeing in urban environments
4. enhancing understanding of interconnected urban health and wellbeing issues across multiple sectors and disciplines, by, e.g. creating a platform for information exchange and knowledge co-production, or pilot projects (in collaboration with ICSU or universities)
5. mediating science and technology alliances for implementing smart technology solutions for urban health and wellbeing
6. Influencing the international policy agenda to encompass themes on urban health and wellbeing initiatives, e.g. Habitat III, Sustainable Development Goals (SDGs).

An overall indicator of the success of the programme will be the extent to which the systems approach, its concepts and methods are known, implemented and become anchored in research, action and urban decision-making. The programme will engage in partnerships with scientists and scientific organisations, in particular with ICSU partner organisations and programmes, for example by establishing research networks, networks of engaged citizens and initiatives, by organising science-policy dialogues but also by connecting to projects and initiative in civil society, business and industry, connecting them to the respective science communities and officially recognising projects which contribute to enhancing the health and wellbeing of urban residents.

Initiatives addressing, for example healthy urban food systems, health education, urban waste management and pollution control, smart healthcare, urban green space, in a systemic way, may seek recognition or endorsements from the programme. For that purpose, the programme will define endorsement criteria and set up a committee which evaluates applications for endorsement. Endorsed projects will be promoted on the programme webpage and disseminated in appropriate ways, in order to improve their visibility and public awareness, in agreement with the parties seeking endorsement.

The programme will deliver outputs which result from its activities with science and society (Table 3). The programme can be regarded as a clearinghouse mechanism on systems science for urban health and wellbeing. By its activities, the programme will encourage ongoing and future science programmes of ICSU to adopt a systems approach which addresses scientific and societal concerns.

Table 3 Examples of potential implementation outputs

Science collaborations	• Knowledge networks for urban health and wellbeing • Publication of scientific studies which provide evidence of health and wellbeing benefits from taking systems approaches • Capacity building, education and training courses and material for taking systems approach to the study of urban health and wellbeing • Curriculum development for systems approaches to urban health and wellbeing • Science-policy dialogues on pressing issues of urban health and wellbeing • Science and technology development of systems solutions for (smart) urban health and wellbeing in collaboration with industry, business and city governments
Private sector partnerships	• Smart urban infrastructure development for health and wellbeing in cities • Identifying innovative technologies for urban health • Partnering with urban planners, property and land developers
Society partnerships	• Civil society and non-governmental organisations engage in projects and activities which demonstrate their views on what is "my healthy city" • Schools educate and engage in urban health and wellbeing activities • Movements and initiatives, such as "Right to the City" constructively engage in urban • Public urban health projects call for participation and public engagement • Citizens participate in urban decision making processes which affect their health and wellbeing • Citizens and urban neighborhoods co-produce public goods and services for urban health and wellbeing • Employers public authorities create opportunity spaces for

9 The Kind of City People Want

Change towards healthy lives in healthy cities requires knowledge of the "kind of problem a city is" as well as "the type of city people want to live in". Given people want to live healthy lives in healthy urban environments, healthy cities are context and culture specific. Despite our knowledge of cities as complex systems providing a range of goods and services, people need to have ideas of how to create their urban environment, what is possible and achievable. Ideas are often inspired by nature and the environments external to human minds. Making cities attractive, viable and healthy is a process of reflecting what people want in urban form, structure, social interaction patterns and urban institutions. Both, what people want and the urban environments they create are interrelated. Seen that way, people and their urban environment form a self-creating (autopoietic) system. Clark (2008) wrote:

> It matters that we recognize the very large extent to which individual human thought and reason are not activities that occur solely in the brain … it drives home the degree to which environmental engineering is also self-engineering. In building our physical and social worlds, we build (or rather, we massively reconfigure) our minds and our capacities of thought and reason.

Recognizing the kind of city people want shifts attention from cities as complex adaptive systems which can be analysed scientifically in a positivist sense by applying scientific tools, to cities as self-creating or self-organizing systems with urban agents (people) who have cognitive abilities. The idea of cities as self-organized systems (Portugali 2000) bridges the hard and soft sciences or the "two cultures" (Snow 1998), like we have attempted to do so with our 'systems approach'. Furthermore, the systems approach and the perception of the city as a complex and self-organised system can help us guide actions in two ways (Portugali 2000) which are also congruent with the systems approach we have proposed:

1. "Taming the city": understand complexity and learn how to identify control parameters, simulate, model and predict urban change for the health and well-being of its residents.
2. "Co-creating the city": participative urban planning and agency.

Conventional ways of working towards the goal of healthy cities by applying expert knowledge within disciplinary domains will not bring about that desired change. Change towards healthy lives in healthy living city systems requires inter- and transdisciplinary, interconnected, integrated and inclusive approaches—approaches we have defined as systems approaches and elsewhere have been referred to as eco-social approaches.

The 2016 United Nations conference on housing and sustainable urban development will agree on the New Urban Agenda and it will set a new global strategy around urbanization for the next two decades. In preparation to conference, the United Nations International Institute of Global Health asked twenty-five urban thought leaders what it takes to create healthy cities. Their thoughts are reflected in the "Principles for healthy and sustainable places" (Box 2).

Box 2: Principles for healthy and sustainable places

1. Ignite the senses: Humans (like animals) are intimately connected to place through the senses—hearing, touch, smell, taste and sight. Beautiful, vibrant and culturally distinct places activate the senses bringing joy, happiness, creativity and generosity of spirit.
2. Build for health and wellbeing: Human and ecosystem health are center stage in good place design. Physical activity, mental health, social connection and healthy eating are fostered through natural and built environments that are designed to connect, respect and protect.
3. Collaborate for equity and access: Justice and equity are longstanding beacons for human health and development in human settlements. Burdens of climate change, unsustainable development and inadequate place design should not be carried by the most vulnerable citizens, cities and countries.
4. Exude openness and flexibility: Prerequisites for innovation and change are human health and a willingness to view things afresh, differently. Lateral, creative thinking transforms challenges into new opportunities for health and sustainability.
5. Embrace diversity and complexity: Natural, economic and cultural diversity make for complex but equally interesting and resilient places. Welcome chaos by engaging with diverse ideas, cultures and approaches to realise healthy, fair and sustainable places.
6. Seek balance and harmony: Be inspired by the laws of nature. Understand the city as a complex human created ecosystem nested within natural ecosystems. Bring together the physical, economic and social elements of places in ways that enhance people-people, and people-ecosystem harmony.
7. Engage citizens dynamically: Citizens are not spectators. Mobilise community around big questions and strive for local answers. Community-led processes help places become stronger and thrive.
8. Enact participatory governance: Good governance is key to the future of human settlements. For many people, participative, accountable and effective systems of governance support place-making ambitions.
9. Privilege integration and holistic approaches: Social, environmental and economicgoals are integrated and indivisible. Siloed, narrow approaches are trumped by partnership-based, multi-level ways of doing business.
10. Contemplate the paradoxes: Consuming less to have more. Invisible in a city of millions. Cars thrive, people and planet suffer. Notice paradoxes that guide responses to modern threats; climate change, economic instability and health inequity

Source: United Nations University (2015). Principles for Healthy and Sustainable Places. Kuching. UNU-IIGH, www.thriveurban.info.

Further, a statement was commissioned by the United Nations International Institute for Global Health (UNU-IIGH), also in the preparation for the United Nations conference HABITAT III, which draws on the insights of the Wolrd Urban Campaign Thinkers Campus held Jan 24–27, 2016 in Kuching, Malaysia. The statement conveys that

1. the health of people and the planet are closely interdependent. People cannot thrive without the Earth's ecosystem goods and services. Therefore, cities need to operate within the limits of Earth's ecological boundaries.
2. the core business of cities should be the people's physical, mental and social wellbeing and that successful cities focus on quality of life indicators instead of merely economic growth measures. Healthy cities are socially inclusive as well as ecologically sustainable. They support innovation, creativity and lifelong education in all their citizens. They also support physical activity, mental wellbeing and social connections. By co-producing safe, clean and beautiful environments; restoring areas of natural and cultural heritage; creating great public places; and enabling active transportation, mobility, accessibility and contact with nature, healthy cities are created.
3. people are at the heart of urban governance and that people adopt governance for health. That governance is participatory, socially just, ecological sustainable.
4. the healthy city goals cannot be achieved with business as usual and any one scientific method. The same tools and approaches which have been applied when creating current urban health problems cannot be used to solve them. New tools and approaches are required for building healthy cities. In particular, eco-social and systems approaches which also make use of indigenous and local knowledge, participatory community education, creative arts and technology. They are all considered important for sustainable urban change.

References

Alberti M (2008) Advances in urban ecology. Integrating humans and ecological processes in urban ecosystems. Springer, New York

Alberti M (2010) Maintaining ecological integrity and sustaining ecosystem function in urban areas. Curr Opin Environ Sustainabil 2:178–184

Albeverio S, Andrey D, Giordano P et al (2007) The dynamics of complex urban systems an interdisciplinary approach. Physica Verlag (Springer), Heidelberg

Albrecht K (ed) (2003) The power of minds at work: organizational intelligence in action, vol 42. American Management Association, New York

Alkire S (undated) The capability approach and human development. http://www.ophi.org.uk/wp-content/uploads/OPHI-HDCA-SS11-Intro-to-the-Capability-Approach-SA.pdf. 24 Nov 2015

Andris C, Bettencourt LMA (2014) Development, information and social connectivity in Côte d'Ivoire. Infrastruct Complex 1:1

Angel S, Sheppard S, Civco D (2005) The dynamics of global urban expansion: Transport and Urban Development Department. World Bank, Washington, DC

Ashby WR (1960) Design for a brain: the origin for adaptive behavior. John Wiley, New York

Atlee T (2006) Co-intelligence, collective intelligence, and conscious evolution. In: Tovay M (ed) Collective intelligence: creating a prosperous world at peace. Earth Intelligence Network, pp 1–4

Axelrod R, Cohen MD (1999) Harnessing complexity: organizational implications of a scientific frontier. The Free Press, New York

Ayres RU, Simonis UE (eds) (1994) Industrial metabolsim: restructuring for sustainable development. United Nations University Press, New York

Baker JL (2008) Urban poverty: a global view. Urban Papers, No. UP-5. World Bank, Washington, DC

Batty M (2003) The emergence of cities: complexity and urban dynamics. Working Paper 64, Centre for Advanced Spatial Analysis, University College London

Batty M, Marshall S (2012) Complexity theories of cities: achievements, cristicism and potentials. In: Portugali J, Meyer H, Stolk E et al (eds) Complexity theories of cities have come of age: an overview with implications to urban planning and design. Springer, Berlin

Beard JR, Petitot C (2011) Ageing and urbanization: can cities be designed to foster active ageing? Public Health Rev 32(2):427–450

Beckmann V, Padmanabhan M (2009) Institutions and sustainability, political economy of agriculture and the environment—essays in honour of Konrad Hagedorn. Springer, Berlin

Bellman R (1964) Control theory. Sci Am 211:186–200

Béné C, Godfrey Wood R, Newsham A et al (2012a) Resilience: new utopia or new tyranny? Reflection about the potentials and limits of the concept of resilience in relation to vulnerability reduction programmes. Working Paper 405, Institute of Development Studies, Brighton, UK

Béné C, Godfrey Wood R, Newsham A et al (2012b) Resilience: new utopia or new tyranny? Reflection about the potentials and limits of the concept of resilience in relation to vulnerability reduction programmes. Working Paper 405. Institute of Development Studies, Brighton

Bertaud A (2014a) Housing affordability: top-down design and spontaneous order. www.nyudri.org/wp-cont,ent/uploads/2011/10/DRIWP96.pdf. 6 June 2015

Bertaud A (2014b) Housing affordability: top-down design and spontaneous order. http://www.nyudri.org/wp-content/uploads/2011/10/DRIWP96.pdf. 10 Jan 2015

Bertaud A (2014c) Working Paper No. 7 the formation of urban spatial structures: markets vs design. http://marroninstitute.nyu.edu/uploads/content/The_Formation_of_Urban_Spatial_Structures.pdf. 1 Oct 2015

Bettencourt L (2013) The kind of problem the city is. SFI Working Paper, Santa Fe Institute, USA

Bettencourt L, Brelsford CH (2015) Industrial ecology: the view from complex systems. J Ind Ecol 19(2):195

Bettencourt LMA, Lobo J, Strumsky D (2007) Invention in the city: Increasing returns to patenting as a scaling function of metropolitan size. Res Policy 36:107–120

Bettencourt L, Hand J, Lobo J (2014) The science and practice of urban planning in slums. Paper prepared for presentation at the "2014 World bank conference on land and poverty". The World Bank, Washington DC

Biermann F, Boas I (2010) Preparing for a warmer world: towards a global governance system to protect climate refugees. Global Environ Politics 10(1):60–88

Brondizio E, Gatzweiler F (2010) Socio-cultural context of ecosystem and biodiversity valuation. In: Kumar P (ed) The economics of ecosystems and biodiversity: the ecological and economic foundations (TEEB). Earthscan, London

Brown H, Proust K, Spickett J et al (2011) The potential role of health impact assessment in tackling the complexity of climate change adaptation for health. Health Promot J Australia 22(s1):s48–s53

Brown RR, Deletic A, Wong THF (2015) Interdisciplinarity: how to catalyse collaboration. Nature 525:315–317. http://www.nature.com/news/interdisciplinarity-how-to-catalyse-collaboration-1.18343

Butler CD, Harley D (2010) Primary, secondary and tertiary effects of the eco-climate crisis: the medical response. Postgrad Med J 86:230–234

Caldwell J, Caldwell B (2002) Poverty and mortality in the context of economic growth and urbanization. Asia Pac Popul J 17:49–66
Clark DA (2005) Sen's capability approach and the many spaces of human well-being. J Dev Stud 41(8):1339–1368
Clark A (2008) Supersizing the mind: embodiment, action, and cognitive extension. Oxford University Press, New York
Coast E (2011) Health and the urban poor: LSE cities. https://lsecities.net/media/objects/articles/health-and-the-urban-poor/en-gb/. 30 Aug 2015
Costanza R (1996) Ecological economics: reintegrating the study of humans and nature. Ecol Appl 6:978–990
Daly HE (1992) Steady-state economics: concepts, questions, policies. Gaia 6:333–338
Daly HE (1993) Sustainable growth: an impossibility theorem. Daly HE, Townsend KN (eds). Valuing the earth. MIT Press, Cambridge
Damasio A (2010) Self comes to mind: constructing the conscious brain. Pantheon Books, New York
Damasio A (2011) Neural basis of emotions. Scholarpedia 6(3):1804. doi:10.4249/scholarpedia.1804
Davidson DJ (2010) The applicability of the concept of resilience to social systems: some sources of optimism and nagging doubts. Soc Nat Resour 23:1135–1149
Dendrinos D, Mullaly H (1985) Urban evolution: studies in the mathematical ecology of cities. Oxford University Press, New York
Easterly W (2008) Institutions: top down or bottom up? Am Econ Rev 98(2):95–99
Ellsberg D (1961) Risk, ambiguity and savage axioms. Quart J Econ 75:643–679
Fischer-Kowalski M (1998) Society's metabolism. J Ind Ecol 2:61–78. doi:10.1162/jiec.1998.2.1.61
Fong WK, Matsumoto H, Lun YF (2009) Application of system dynamics model as decision making tool in urban planning process toward stabilizing carbon dioxide emissions from cities. Build Environ 44(7):1528–1537
Freudenberg N, Klitzman S, Saegert S (2009) Urban health and society: interdisciplinary approaches to research and practice. Wiley, San Francisco
Funtowicz S, Ravetz J (1990) How to cope with uncertainty? Uncertainty and quality in science for policy. Springer, Netherlands
Funtowicz S, Ravetz J (1993) Science for the post-normal age. Futures 25:735–755
Funtowicz S, Ravetz J (2003) Post-normal science. In: Internet encyclopaedia of ecological economics. In: International Society for Ecological Economics
Gatzweiler FW (2013) The meaning of value and institutional change for an economics of living systems. Habilitationsschrift, Humboldt University of Berlin, Berlin
Geddes P (1949) Cities in evolution. Williams and Norgate, London
Glimcher P (2003) Decisions, uncertainty and the brain: the science of neuroeconomics. MIT Press, Cambridge
Glimcher PW, Camerer CF, Fehr E et al (eds) (2009) Neuroeconomics: decision making and the brain. Elsevier, Amsterdam
Gómez-Baggethun E, Gren A, Barton ND et al (2013) Urban ecosystem services. In: Elmqvist T, Fragkias M, Goodness J et al (eds) Urbanization, biodiversity and ecosystem services: challenges and opportunities. Springer Netherlands, Dordrecht. doi:10.1007/978-94-007-7088-1
Grimm NB, Baker LJ, Hope D (2011) An ecosystem approach to understanding cities: familiar foundations and uncharted frontiers. In: Berkowitz AR, Nilon CH, Hollweg KS (eds) Understanding urban ecosystems: a new frontier for science and education. Springer, New York
Gunderson LH, Holling CS (2002) Panarchy: understanding transformations in human and natural systems. Island Press, Washington
Gurr JM, Walloth C (2014) Introduction: towards a transdisciplinary understanding of complex urban systems. In: Walloth C, Gurr JM, Schmidt JA (eds) Understanding complex urban systems: multidisciplinary approaches to modeling. Springer, Cham

Hämäläinen RP, Saarinen E (eds) (2004) Systems intelligence: discovering a hidden competence in human action and organizational life. Helsinki University of Technology, Systems Analysis Laboratory Research Reports, vol 9

Hancock T (undated) WHO healthy cities project: promoting health in the urban context. WHO Healthy Cities Papers No. 1, vol 24

Hancock T, Duhl L (1988) WHO, healthy cities project: a guide to assessing healthy cities. FADL Publishers, Copenhagen

Hancock T, Spady DW, Soskolne CL (2015) Global change and public health: addressing the ecological determinants of health. http://www.cpha.ca/uploads/policy/edh-brief.pdf. 1 Oct 2015

Harpham T, Tanner M (1995) Features and determinants of urban health status. In: Harpham T, Tanner M (eds) Urban health in developing countries. Earthscan Publications Limited, London, pp 35–47

Harris AS, Keil R (eds) (2008) Networked disease: emerging infections in the global city. Blackwell Publishing Ltd., UK

Hayek FA (1945) The use of knowledge in society. Am Econ Rev 35(4):519–530

Heymann DL, Barakamfitiye D, Szczeniowski M et al (1999) Ebola hemorrhagic fever: lessons from Kikwit, Democratic Republic of the Congo. J Infect Dis 179(Suppl 1):S283–S286

Holland JH (1998) Emergence: from chaos to order. Addison-Wesley, Redwood City

Holston J (2008) Insurgent citizenship. Princeton University Press, Princeton

ICSU (2011a) ICSU Strategic Plan II, 2012–2017. International Council for Science, Paris

ICSU (2011b). Report of the ICSU planning group on health and wellbeing in the changing urban environment: a systems analysis approach. International Council for Science, Paris

Jacobs J (1961) The death and life of great American cities. Random House, New York

Janssen M, Scheffer M (2004) Overexploitation of renewable resources by ancient societies and the role of sunk-cost effects. Ecol Soc 9(1):2

Kahn ME (2006) The benefits of sprawl. [2015-10-20]. http://greeneconomics.blogspot.de/2006/03/benefits-of-sprawl.html. 20 Oct 2015

Kennedy CA, Cuddihy J, Yan E (2007) The changing metabolism of cities. J Ind Ecol 11:43–59

Keysers C, Cohen J, Donald M et al (2008) Explicit and implicit strategies in decision making. In: Engel C, Singer W (eds) Better than conscious? implications for performance and institutional analysis. Strungman Forum Report. MIT Press, Cambridge

Knight R, Buhler B (2015) Follow your gut: the enormous impact of tiny microbes. Simon & Schuster TED Books, New York

Lawrence R (2015) Advances in transdisciplinarity: epistemologies, methodologies and processes. Futures 65:1–9

Leach M, Scoones I, Stirling A (2010) Governing epidemics in an age of complexity: narratives, politics and pathways to sustainability. Glob Environ Change 20:369–377

Lentz DL, Dunning NP, Scarborough VL (2015) Tikal, paleoecology of an ancient Maya City. Cambridge University Press, New York

Lovelock JE (1972) Gaia as seen through the atmosphere. Atmos Environ 6(8):579–580

Maahsen-Milan A, Pellegrino M, Oliva L et al (2013) Urban architecture as connective-collective intelligence. Which spaces of interaction? Sustainability 5:2928–2943. doi:10.3390/su5072928

Malone TW (2006) What is collective intelligence and what will we do about it?. Tovay M (ed) Collective intelligence: creating a prosperous world at peace. earth intelligence network

Mattern S (2014) Interfacing urban intelligence. Places J. https://placesjournal.org/article/interfacing-urban-intelligence/. 6 Oct 2015

McMichael AJ (2013) Globalization, climate change, and human health. N Engl J Med 368: 1335–1343. doi:10.1056/NEJMra1109341

McMichael AJ, Butler CD (2011) Promoting global population health while constraining the environmental footprint. Annu Rev Public Health 32:179–197

Mcphearson T, Hamstead ZA, Kremer P (2014) Urban ecosystem services for resilience planning and management in New York City. Ambio 43(4):502–515. doi:10.1007/s13280-014-0509-8. 9 Aug 2015

Mehta L, Leach M, Newell P et al (1999) Exploring understandings of institutions an uncertainty: new directions in natural resource management. IDS Discussion Paper 372, Environment Group, Institute of Development Studies, University of Sussex, Brighton, BN1 9RE
Ming HSU, Bhatt M, Adolphs R et al (2005) Neural systems responding to degrees of uncertainty in human decision-making. Science 310(5754):1680–1683
Mitchell D (2003) The right to the city: social justice and the fight for public space. The Guilford Press, New York
Napier AD, Ancarno C, Butler B (2014) Culture and health. Lancet 384(9954):1607–1639. doi:10.1016/S0140-6
Naqvi N, Shiv B, Bechara A (2006) The role of emotion in decision making a cognitive neuroscience perspective. Curr Dir Psychol Sci 15(5):260–264
North DC (1990) Institutions, institutional change, and economic performance. Cambridge University Press, New York
Odum EP (1971) Fundamentals of ecology (3rd ed). Saunders, Philadephia
OECD (2014) The cost of air pollution: health impacts of road transport. OECD Publishing, Paris. doi:10.1787/9789264210448-en
Olsson L, Jerneck A, Thoren H et al (2015) Why resilience is unappealing to social science: theoretical and empirical investigations of the scientific use of resilience. Sci Adv 22:E1400217
Ostrom V (1983) Freedom and organization. Working Paper W83-16, Workshop in Political Theory and Policy analysis, Indiana University, IN, USA
Periyakoil VS (2007) Taming wicked problems in modern health care systems. J Palliat Med 10(3):658–659
Pickett STA, Burch WR, Dalton SE et al (1997) A conceptual framework for the study of human ecosystems in urban areas. Urb Ecosyst 1:185–199
Portugali J (ed) (2000) Self-organization and the city. Springer, Berlin
Proust K, Newell B (2016) Urban sub-system influence diagrams, designing health places: systems thinking and the SDGs. United Nations University-Int Inst Global Health, Urban Thinkers Campus, Kuching, Sarawak
Puppim de Oliveira J, Doll CN, Siri J et al (2015) Urban governance and the systems approaches to health-environment co-benefits in cities. Cad. Saúde Pública, Rio de Janeiro 31(Sup): S25–S38. 24 Dec 2015. http://www.scielo.br/pdf/csp/v31s1/0102-311X-csp-31-s1-0025.pdf. [Undated]
Rees W (1995) Achieving sustainability: reform or transformation? J Plann Lit 9:343–361
Rees W (2003) Understanding urban ecosystems: an ecological economics perspective. In: Berkowitz AR, Nilon CH, Hollweg KS (eds) Understanding urban ecosystems. Springer, New York, p 124
Reid WV, Mooney HA, Cropper A (2005) Millennium ecosystem assessment. Ecosystems and human well-being: synthesis. Island Press, Washington, DC
Rockstrom J, Steffen W, Noone K et al (2009) A safe operating space for humanity. Nature 461:472–475
Rook G (2013) Regulation of the immune system by biodiversity from the natural environment: an ecosystem service essential to health. PNAS 110:18360–18367
Rydin Y, Bleahu A, Davies M et al (2012) Shaping cities for health: complexity and the planning of urban environments in the 21st century. Lancet 379:2079–2108
Sanders TI (2008) Complex systems thinking and new urbanism. Haas T (ed) New urbanism and beyond: designing cities for the future. Rizzoli, New York
Satterthwaite D, Huq S, Reid H et al (2009) Adapting to climate change in urban areas: the possibilities and constraints in low- and middle-income nations. In: Bicknell J, Dodman D, Satterthwaite D (eds) Adapting cities to climate change: understanding and addressing the development challenges. Earthscan, London
Schelling T (1960) The strategy of conflict. Harvard University Press, Cambridge
Sen A (1985) Wellbeing, agency and freedom: the dewey lectures 1984. J Philos 82(4):203
Sen A (1999) Development as freedom. Harvard University Press, Cambridge

Seto KC, Dhakal S (2014) Chapter 12: Human settlements, infrastructure, and spatial planning. In: Edenhofer O, Pichs-Madruga R, Sokona Y et al (eds) Climate change 2014: mitigation of climate change. Contribution of Working Group III to the fifth assessment report of the intergovernmental panel on climate change. Cambridge University Press, Cambridge. http://www.mitigation2014.org, 14 Oct 2015

Snow CP (1998) The two cultures. Cambridge University Press, Cambridge

Stanovich KE, West RF (2000) Individual differences in reasoning: implications for the rationality debate. Behav Brain Sci 23:645–665

Stirling A (2006) Uncertainty, precaution and sustainability: Towards more reflective governance of technology. In: Voss J, Kemp R (eds) Sustainability and reflexive governance. Edward Elgar Publishing, Cheltenham, pp 25–272

Tannert C, Elvers HD, Jamdrig B (2007) Ethics of uncertainty. European molecular biology organization. EMBO Rep 8(10):892–896

TEEB (2010) The economics of ecosystems and biodiversity. Ecological and Economic Foundations, Earthscan, London

TEEB (2011) The economics of ecosystems and biodiversity. TEEB manual for cities: Ecosystem Services in Urban Management. www.teebweb.org. 10 Aug 2015

Tovay M (ed) (2006) Collective intelligence: creating a prosperous world at peace. Earth Intelligence Network

Uhl-Bien M, Marion R, Mckelvey B (2007) Complexity leadership theory: shifting leadership from the industrial age to the knowledge era. Leadership Institute Faculty Publications

UN-Habitat (2008) The state of African cities. www.unhabitat.org. 22 Aug 2016

UN-Habitat. State of the world's cities 2006/2007. http://www.unhabitat.org/pmss/listItemDetails. 20 Aug 2006

United Nations (2013) Report on the world's social situation. http://www.un.org/esa/socdev/documents/reports/InequalityMatters.pdf. Accessed 6 Oct 2015

United Nations, Department of Economic and Social Affairs (2014) World urbanization prospects: the 2014 revision. United Nations, New York

Vatn A (2009) An institutional analysis of methods for environmental appraisal. Ecol Econ 68:2207–2215

Vester F (2007) The art of interconnected thinking: tools and concepts for a new approach tackling complexity. MCB Publishing House (Management, Cybernetics, Bionics), Munich

Vlahov D, Freudenberg N, Proietti F et al (2007) Urban as a determinant of health. J Urb Health: Bull NY Acad Med 84(Suppl 1):16–26. doi:10.1007/s11524-007-9169-3; http://visibleearth.nasa.gov/view.php?id=55167; http://earthobservatory.nasa.gov/. 2 Oct 2015

Walker B, Meyers JA (2004) Thresholds in ecological and social-ecological systems: a developing database. Ecol Soc 9(2):3

Walloth C, Gurr JM, Schmidt JA (eds) (2011) Understanding complex urban systems: multidisciplinary approaches to modeling. Springer, Cham

Weick K, Sutcliffe K (2001) Managing the unexpected-assuring high performance in an age of complexity. Wiley, San Francisco

Whitmee S, Haines A, Beyrer C et al (2015) Safeguarding human health in the anthropocene epoch: report of the Rockefeller Foundation—Lancet Commission on planetary health. Lancet, 10 Aug 2015. doi:10.1016/S0140-6736(15)60901-1. 6 July 2015

WHO (2015) World report on ageing and health. World Health Organization, Geneva

Endnotes

[1]World Cities Report 2016. Accessed Sep 1, 2016 @ http://wcr.unhabitat.org/quick-facts/.

[2]The UK's chief scientific adviser Sir Mark Walport at the Royal Society's Future of Scholarly Scientific Communication conference, held in London on 5–6 May 2015, suggested scientists to communicate their results through "evolving manuscripts" that are updated continually over a working life; or 'citizen science'—teams of scientists and lay people who make observations, contribute to data collection and help classify data. See: http://www.scientificamerican.com/citizen-science/?category=health-and-medicine.

[3]See the nature journal's special issue on interdisciplinarity: http://www.nature.com/news/interdisciplinarity-1.18295, Nature 2015 (525), doi:10.1038/525289b.

[4]United Nations human settlement programme. http://unhabitat.org/.

[5]See: http://www.globalgoals.org/global-goals/sustainable-cities-and-communities/.

[6]WDS—World Data System, https://www.icsu-wds.org/.

[7]See: CoData—Committee on Data for Science and Technology, http://www.codata.org/about-codata/our-mission.

[8]The Intergovernmental Panel on Climate Change (IPCC) estimates that in 2010, urban areas accounted for 67–76% of global energy use and 71–76% of global CO2 emissions from final energy use (Seto and Dhakal 2014).See also: The new climate economy report "Better Growth, Better Climate" Chap. 2: Cities.

[9]17 Sustainable Development Goals: http://www.un.org/sustainabledevelopment/.

[10]Accessed Oct 2, 2015 at: http://visibleearth.nasa.gov/view.php?id=55167; http://earthobservatory.nasa.gov/.

[11]Also referred to as complex adaptive systems. Sanders (2008) listed the following characteristics of cities as complex adaptive systems:

1. Diversity among the components; heterogeneous parts or "agents;" sources of novelty in the system. Includes some sort of natural selection processes within agent groups that ensure ongoing evolution, regeneration, and adaptation.
2. Nonlinear interactions; widespread information flow and feedback loops.
3. Self-organization; results from attractors in the system, and from adaptation to changes in the larger environment and other agents.

F.W. Gatzweiler et al., *Advancing Health and Wellbeing in the Changing Urban Environment*, Urban Health and Wellbeing, DOI 10.1007/978-981-10-3364-3

4. Local information processing; local interactions among autonomous agents. Typically agents "see" only their part of the system and act locally; no global control.
5. Emergence; exhibits unpredictable global behavior or patterns; spontaneous order emerges from local system interactions.
6. Adaptation; open and responsive to changes to the larger environment or context and to other agents in the system; continuously processing, learning, and incorporating new information; making boundaries hard to define.
7. Organization across multiple scales; agents in the system organized into groups or hierarchies of some sort, which influence how the system evolves over time.
8. Sensitivity to changes in initial conditions; small changes can create big results at some point in the future.
9. Non-equilibrium; most interesting behavior/creativity found at the "edge of chaos;" healthy systems operate in a dynamic state somewhere between the extremes of order and disorder, making it easier for them to adapt to changing conditions.
10. Best understood by observing the behavior—activities, processes, adaptation—of the whole system over time; qualitative descriptions and understanding versus quantitative descriptions alone.

[12]Here we define megacities as the largest cities of their time, in terms of population size.

[13]For example the ancient Mayan city of Tikal in Guatemala or Angkor in Cambodia (Lentz et al. 2015).

[14]UN-HABITAT. The State of African Cities. 2014. http://unhabitat.org/books/state-of-african-cities-2014-re-imagining-sustainable-urban-transitions/state-of-african-cities-2014-re-imagining-sustainable-urban-transitions/.

[15]The New Climate Economy Report: Chapter 2: Cities; www.newclimateeconomy.net.

[16]This view relates to the Gaia Theory developed by Lovelock (1972), which describes a productive unity between scientific understandings of planet Earth as a living system with cultural understandings of human society as a seamless continuum of that system.

[17]Credit: NASA Earth Observatory image by Robert Simmon, using Suomi NPP VIIRS data provided courtesy of Chris Elvidge (NOAA National Geophysical Data Center). Suomi NPP is the result of a partnership between NASA, NOAA, and the Department of Defense. Caption by Mike Carlowicz.

[18]Governance is here defined as socio-political and institutional processes and strategies, which are based on different ideas, knowledge and power.

[19]For example: The cost of the health impact of air pollution in OECD countries (including deaths and illness) was about USD 1.7 trillion in 2010 (OECD 2014).

[20]Physical spaces are areas or places. Social or organizational spaces are positions or locations in social networks and organizations. Institutional spaces are opportunities or incentives provided by sets of rules. Economic spaces can be

areas in which a business has a competitive advantage. Ecological spaces can be physical locations of species competing for nutrition or sunlight or niches occupied by genetic adaptation or mutation. The idea that multidimensional urban spaces interact in complex systems was pioneered by Jacobs (1961). See also the "spaces of wellbeing" by Clark (2005).

[21]Modified from Leach et al. (2010).

[22]See issues paper on Smart Cities and Infrastructure by UNCTAD, accessed Jan 2016: http://unctad.org/meetings/en/SessionalDocuments/CSTD_2015_Issuespaper_Theme1_SmartCitiesandInfra_en.pdf.

[23]Intelligent cities. Routes to a Sustainable, Efficient and Livable City, http://www.intelligent-cities.net/Glossar.html.

[24]Adelaide Statement on Health in All Policies, http://www.who.int/social_determinants/hiap_statement_who _sa_final.pdf.

[25]See the London school of Economics (LSE) Cities projects on "Urban Uncertainty" and "New Urban Governance".

[26]For example, Ashby's rule of requisite variety which says that any regulative system needs as much variety in the actions that it can take as exists in the system it is regulating (Ashby 1960).

[27]From a neuroscientific perspective order (relationships of the same type) in neural networks of the brain allows human beings to make decisions in situations of complexity and uncertainty. Neurobiological modes of decision-making vary between analytical and explorative modes. The latter are related to socio-emotional neural regions of the brain and, when made use of, improve the quality of decision-making in complex and uncertain decision making situations. This behavioral response to uncertainty emphasizes the importance of failure as a signal for structural renewal. Stanowich and West (2000) differentiate two systems of response to uncertainty: rule-based or analytical versus explorative or experimental decision making processes. Both types of decision-making are associated with different brain regions and cannot be strictly separated (Keysers et al. 2008). The analytic type involves the prefrontal and parietal regions of the neocortex and allows for planning, cognitive control and self-regulation. The experimental type often involves strong feelings and the activation of the socio-emotional neural networks. They are located in the limbic and paralimbic regions of the brain and are evolutionary much older than the neocortical regions. (…). Broadly, the quality of decision-making improves in complex decision-making situations characterized by uncertainty, when making use of the socio-emotional, explorative neural system. Damasio (2010, 2011) researched the interplay of both behaviors, especially with patients who are intelligent and analytical thinkers but unable to perform in (everyday life) situations of uncertainty. These patients had an injury in the forebrain and were unable to make use of emotions for decision making." (Gatzweiler 2013).

[28]Tannert et al. (2007) define different types of uncertainty ranging from "not possible to know" to "not wanting to know".

[29]Two systems of response to uncertainty can be distinguished: rule-based or analytical versus explorative or experimental decision making processes. Both types of decision-making are associated with different brain regions and cannot be strictly separated. Broadly, the quality of decision-making improves in complex decision-making situations characterized by uncertainty, when making use of the socio-emotional, explorative neural system.

[30]For example the Chinese solar rule of the 1950, in which China established a regulation requiring that at least one room in each apartment receive a minimum of one hour of sunshine on the day of the winter solstice, December 21 (Bertaud 2014).

[31]For example the "kampongs" in the vicinity of Surabaya city, Indonesia.

[32]People's health and wellbeing can be improved in various dimensions and respective spaces. Social, ecological, economic dimensions and respective physical and temporal and conceptual spaces. For example, making trade-offs between present versus future consumption and savings, or living in crowded versus greener areas.

[33]Communication barriers between science and society refer to the diversity of languages, terminology and concepts used and understood, as well as the media and opportunities of communication, e.g. workshops, conferences, webinars, fora, roundtables.

[34]United Nations University "What does good governance mean?", http://unu.edu/publications/articles/what-does-good-governance-mean.html; See also: International Fund for Agricultural Development (IFAD) Executive Board, Sixty-Seventh Session, Rome, 8–9 September 1999, Good Governance: an Overview, http://www.ifad.org/gbdocs/eb/67/e/EB-99-67-INF-4.pdf

The Cotonou Partnership Agreement of 2000, Article 9.3, defined good governance: ""In the context of a political and institutional environment that upholds human rights, democratic principles and the rule of law, good governance is the transparent and accountable management of human, natural, economic and financial resources for the purposes of equitable and sustainable development."

[35]In this paper Vincent Ostrom defines freedom as "… the capacity to act on the basis of one's own considered judgment." Organization is defined as "a requirement to act in relation to mutually understood rules that enable individuals to pursue opportunities subject to limits." The dilemma is that "organization is necessary to freedom and, at the same time, constitutes a fundamental threat to freedom."

[36]Example taken from Funtowicz and Ravetz (2003): "…passenger transport had traditionally been seen as an essentially straightforward engineering problem of maximising mobility, subject to the constraints of optimising costs and safety. Now transport technologies and policies are strongly influenced by environmental considerations of many sorts, including concerns about sustainability. Moreover, consumer demand for passenger transport depends directly on lifestyles. Just now, Americans want large, gas-guzzling vehicles, and Europeans

want cheap air travel to the sun. In terms of post-normal theory, passenger transport presents the severe systematic uncertainties of climate change, combined with the crucial decision stakes in conceptions of the good life, along with considerations of equity between peoples and generations. The entire population of passenger transport users has effectively become an extended peer community. The success of sustainable transport technologies will depend on the effectiveness of the public's commitment to the values of the global environment. "

[37]Organizational intelligence has been defined by Albrecht (2003) as "syntropy", the "coming together of people, ideas, resources, systems, and leadership in such a way as to fully capitalize on the possibilities of each." (See: Albrecht 2003)

[38]Systems Intelligence has been defined as "…intelligent behaviour in the context of complex systems involving interaction and feedback." (See: Hämäläinen and Saarinen 2004)

[39]Collective intelligence has been defined as "…groups of individuals doing things collectively that seem intelligent."(Malone 2006) Google and Wikipedia are examples. The State of the Future 2010 report defines collective intelligence as "an emergent property from synergies among data/information/knowledge, and experts (…) that continually learns from feedback to produce (…) knowledge for better decisions than these elements acting alone." Atlee (2006) distinguishes collective intelligence from co-intelligence which "…embraces more than collective intelligence (CI), the intelligence of groups. It includes at least multi-modal intelligence, collaborative intelligence, resonant intelligence, universal intelligence, and wisdom (see http://tinyurl.com/2128nh). By itself—and especially without wisdom (embracing the big picture)—collective intelligence, like individual intelligence, can be used in harmful ways, such as building gas chambers and new technologies with disastrous "side effects." Citizen juries, assemblies, consensus conferences, participatory decision making, citizen deliberative councils, … are *actually methods for manifesting consciousness at higher scales of social organization by cognitive synergies.* Those are evolutionary processes of conscious building. The scales range from individual, to family, group, organization, city, state, global intelligence. In different forms of collective intelligence, group intelligence can be more powerful than individual intelligence by seeing a bigger, more complete picture and overcoming blind spots—very much depending on how the collective process itself is designed. See also the State of the Future Reports by the Millennium Project. The Millennium Project was founded in 1996 after a three-year feasibility study with the United Nations University, Smithsonian Institution, Futures Group International, and the American Council for the UNU. It is now an independent non-profit global participatory futures research think tank. The Global Futures Intelligence System is a key output of the Millennium Project: http://www.millennium-project.org/index.html.

[40]Some forms of collective intelligence are (1) reflective= people think together using dialogue and deliberation, (2) structure=social systems that support

intelligent behaviors, e.g. the Bill of Rights, which supports free flow of information, (3) evolutionary = ecosystems that co-evolved as manifestations of learning or institutions in academia, (4) informational = The flow of information through communication channels and the widespread gathering and availability of information in databases, universal access to information, e.g. Wikipedia (See: Tovay 2006, p. 10).

[41]Puppim de Oliveira et al. (2015).

[42]Agency is "what a person is free to do and achieve in pursuit of whatever goals or values he or she regards as important" (Sen 1985: 203). Another definition of agency: "agency is the person's ability to act on what they value and have reason to value." (Alkire undated).

[43]This note was developed by the Scientific Committee for the ICSU Urban Health programme.

[44]Major stakeholder groups of the programme are: (1) the academic research community (individual scientists, universities and research organizations), (2) organizations at the science-policy interface (United Nations Organization, Interacademy Medical Panel, Intergovernmental panels, networks, consultancies), (3) research funders, private and public, (4) development organizations, (5) government organizations, especially at municipal level, (6) civil society organizations, (7) media.

[45]World Health Organization (WHO) definition of Health, Preamble to the Constitution of the World Health Organization as adopted by the International Health Conference, New York, JUN 19–22, 1946.

[46]Health and wellbeing of people influence each other in ways which require further investigation, especially for understanding the relations to urban system functions. The overall wellbeing of urban populations, comprises of more than constituents of human health. How well cities function and are able to provide the range of goods and services to their inhabitants, also effects the health and wellbeing of urban populations—but this relationship is not straightforward. A possible relationship between health and wellbeing is that increasing investments in urban health functions, after a threshold, result in diminishing marginal improvements in wellbeing. Even with a well-developed health system a city does not equally provide its functions to all segments of urban society. Taking a systems perspective to understanding health and wellbeing in changing urban environments, therefore requires distinguishing between the condition of the urban system (with regards to its functions) as well as the condition and position of its population in the urban system (referring to their capabilities).

[47]The human microbiome is an extension of the biodiversity of ecosystems and thereby represents a very particular and close link between humans and the microbial diversity of their environments (Rook 2013). That microbiota helps humans digest food, it educates the immune system, regulates inflammation, resists disease and even effects human behaviour (Knight and Buhler 2015). 1,200 different species of bacteria have evolved in symbiosis with the human host over the course of millennia.

[48]Any biotic community interacting with its environment such that the flow and dissipation of energy results in a defined trophic (feeding) structure, the emergence of biodiversity, and characteristic material cycles between the living and non-living components may be considered an ecosystem (Odum 1971).

[49]Humans are obvious also "natural" in the sense of being one species among many on planet earth. This makes the social-ecological or natural vs man-made distinction obsolete. However for conceptual reasons and epistemological tradition we maintain it.

[50]Ayres and Simonis (1994) define industrial metabolism as "… the whole integrated collection of physical processes that convert raw materials and energy, plus labour, into finished products and wastes …".

[51]As compared to autotroph organisms, mainly plants, which are able to fix solar energy to produce biomass.

[52]Urban metabolism can be defined as "the sum total of the technical and socio-economic processes that occur in cities, resulting in growth, production of energy, and elimination of waste" (Kennedy et al. 2007). In human ecology and industrial ecology, a very similar concept of "social metabolism" has been developed (Fischer-Kowalski 1998).

[53]http://www.icsu.org/science-international/accord/open-data-in-a-big-data-world-short.

[54]http://www.scholarpedia.org/article/Self-organization.

Glossary

Adaptive cycle Systems can collapse, rebuild and reorganize after reaching a critical level of connectedness

Agency People's ability to act on what they value and have reason to value being and doing"… what a person is free to do and achieve in pursuit of whatever goals or values he or she regards as important." (Sen 1985: 203)Agency is "what a person is free to do and achieve in pursuit of whatever goals or values he or she regards as important" (Sen 1985: 203).

Capability The capability of a person reflects the alternative combinations of functionings a person can achieve, and from which he or she can choose one collectionThe freedoms which enable agents to realize their valued functionings. The concept of capabilities strongly relates to that of the "enabling institutional environment."—which is a regulating and provisioning function provided by the urban system.Having capabilities enables people to actually make a choice about what they want to be or do and having real opportunities with regards to the functionings they aspire, value and have reason to value.

Civil society The wide array of non-governmental and not-for-profit organizations that have a presence in public life, expressing the interests and values of their members or others, based on ethical, cultural, political, scientific, religious or philanthropic considerations. Civil Society Organizations (CSOs) therefore refer to a wide of array of organizations: community groups, non-governmental organizations (NGOs), labor unions, indigenous groups, charitable organizations, faith-based organizations, professional associations, and foundations (Definition adopted by the World Bank)

Collective intelligence Higher degrees of connectedness and consciousness in a group but also more sophisticated systems of information processing and foresight created by deliberative and participatory research and knowledge co-production. Some forms of collective intelligence are (1) reflective = people think together using dialogue and deliberation, (2) structure = social systems that support intelligent behaviors, e.g. the Bill of Rights, which supports free flow of information, (3) evolutionary = ecosystems that co-evolved as

F.W. Gatzweiler et al., *Advancing Health and Wellbeing in the Changing Urban Environment*, Urban Health and Wellbeing, DOI 10.1007/978-981-10-3364-3

manifestations of learning or institutions in academia, (4) informational = The flow of information through communication channels and the widespread gathering and availability of information in databases, universal access to information (Tovay 2006)

Complex systems "Complex Systems are systems that comprise many interacting parts with the ability to generate a new quality of macroscopic collective behavior the manifestations of which are the spontaneous formation of distinctive temporal, spatial or functional structures." (Walloth et al. 2014)

Co-design Co-design (co-produce, or co-determine) is an participatory approach to actively involve all stakeholders in a process of producing, designing, or determining

Emergence Emergence in a social system is a product of interactions among different agents. The behaviour of one observed agent or a group is the result of others they are interacting with, which again is the result of yet further interactions with other sets of actors or groups. Value can be regarded as an emergent property of social or ecological systems

Functionings People s' lives can be seen as a combination of different doings and beings, which can be called functionings. Functionings are the various things a person may value and have reason to value doing or being. For example, being nourished or riding a bicycleThe capability of a person, in turn is being able to be nourished and being able to ride around. The capability refers to the various alternative combinations of functionings the person can choose to have.

Freedom The real opportunity that people have to accomplish what they valueThe ability to shape one's own destiny as a person and a part of various communities. There are two aspects to freedom: the ability to act on behalf of what matters (agency) and the real opportunity to achieve valued functionings.

Governance Formal and informal policies, budgetary and regulatory decisions made at global, national, city and local scales as well as negotiations over access to power and services that may enhance or detract from urban health and wellbeing

Health The absence of ill health and the presence of mental, social and physical wellbeing

Healthy city One that is continually creating and improving those physical and social environments and expanding those community resources which enable people to mutually support each other in performing all the functions of life and in developing to their maximum potential

Intelligent city The idea of an intelligent, sustainable city includes cities which are green and smart, which integrate core areas (such as energy, mobility, planning, governance and economy) and include cross-sectional topics (such as ICT, innovative finance and public participation)

Risk Situations may happen a probability that can be assessed and calculated with respect to particular outcomes

Self-organization The ability of systems not only to adapt to external environments but also to shape the environment within which they operate. Self-organised systems show circular causality and spontaneous order"The spontaneous often seemingly purposeful formation of spatial, temporal, spatiotemporal structures or functions in systems composed of few or many components. In physics, chemistry and biology self-organization occurs in open systems driven away from thermal equilibrium. The process of self-organization can be found in many other fields also, such as economy, sociology, medicine, technology."[54]

Science The process and organization of how we achieve and communicate knowledge and create collective intelligence

Sustainable Development Goals (SDGs) Internationally agreed goals by more than 150 countries through a United Nations-led process that adopted the 2030 agenda for sustainable development. They include 17 goals and 169 targets

Systems analysis Analytic approaches that aim to understand the complex relationships among components of a system(s) including interrelationships between subsystems of a larger system

Uncertainty Situations which cannot be planned, controlled or known. Tannert et al. (2007) explain that there are different types of uncertainty, which refer to what people cannot know (unknown unknowns), are not capable of knowing, and do not want to know (Galileo effect) (Tannert et al 2007)

Urban density The number of population inhabits a unit of urban area

Urban environment The physical, social, technological and institutional space within which urban activities take place (from last science report) the natural, built and institutional elements that determine the physical, mental and social health and wellbeing of people who live in cities and towns

Urban health systems The complex urban socio- ecological-technological systems providing multiple functions for human health and wellbeing

Urban system functions A kind of choice set from which the human agent aims at choosing to be and do what he values or has reason to value

Urbanization Urbanization is the process by which the fraction of the total population that is urban increases

Wellbeing A situation in which people are free to choose to do and be what they value being and doing

Index

F.W. Gatzweiler et al., *Advancing Health and Wellbeing in the Changing Urban Environment*, Urban Health and Wellbeing, DOI 10.1007/978-981-10-3364-3

FSC
www.fsc.org
MIX
Papier | Fördert
gute Waldnutzung
FSC® C083411

Kolibri 360 GmbH
Ferdinand-Jühlke-Straße 7
99095 Erfurt, Deutschland
produktsicherheit@kolibri360.de